THE HOUSE
OF
SCIENCE

Related Titles of Interest

The Thomas Edison Book of Easy and Incredible Experiments, by the Thomas Alva Edison Foundation

Chemistry for Every Kid: 101 Easy and Incredible Experiments That Really Work, by Janice Pratt VanCleave

Clouds in a Glass of Beer: Simple Experiments in Atmospheric Physics, by Craig Bohren

Nature for the Very Young: A Handbook of Indoor and Outdoor Activities, by Marcia Bowden

The Naturalist's Year: 24 Outdoor Explorations, by Scott Camazine

THE HOUSE
OF
SCIENCE

Philip R. Holzinger

WILEY

Wiley Science Editions

John Wiley & Sons, Inc.

NEW YORK   CHICHESTER   BRISBANE   TORONTO   SINGAPORE

Library of Congress Cataloging-in-Publication Data

Holzinger, Philip R.
  The house of science / Philip R. Holzinger.
    p. cm. — (Wiley science editions)
  Includes bibliographical references.
  ISBN 0-471-50061-5. (pbk.) — ISBN 0-471-51052-1
  1. Science. I. Title. II. Series.
Q161.2.H63 1989
500—dc20   89-37619
                                                CIP

Printed in the United States of America

90 91 10 9 8 7 6 5 4 3 2 1

PREFACE

When we build a house, we must first construct a framework to make it stand. When we pursue a science education, we must likewise construct a framework for organizing what we learn. *The House of Science* was written for the purpose of constructing such a framework. Each chapter provides a foundation for understanding a particular science and questions throughout the chapter stimulate thought and test the comprehension of the material. By answering these questions, students will begin to acquire a working knowledge of that science. They will in effect be constructing their own mental framework—their own "house of science." With this framework, they will be better able to comprehend the role of science in our present-day world. They will also be better equipped to think scientifically.

This book is meant to be used as a science primer. The exercises are designed to be convenient for classroom use as well as for private study; many of them can be done in one or two sittings. Each chapter ends with a summary and answers to the questions in the text.

I would like to thank all the people who helped me with this book, particularly David Sobel and Frank Grazioli, my editors at John Wiley & Sons. Special thanks also go to my father, Joseph Holzinger, for his advice on many aspects of the project and to Jim Muschlitz for reviewing the manuscript in its entirety. I also wish to thank Gil Hoffman, Sally Olson, Bob Price, Don Price, Carl Smith, and Herb Stinson for their help with various aspects of the project. In addition, I would like to thank the following people for graciously reviewing various chapters of the manuscript: Dr. Joseph Gerencher and Dr.

Frank J. Kuserk of Moravian College; and Dr. Edward Beutner, Dr. John J. Farrell, Dr. Philip Sutter, and Steve Sylvester, all of Franklin & Marshall College. Although these reviewers were most helpful, I of course take full responsibility for any errors or omissions in the book.

Finally, I would like to thank Jack Crane for many of the fine illustrations that appear throughout the book, including the cover, and Francis Lawrence, for help with the graphic artwork and layout. Without either of these two individuals, the book would not be what it is.

CONTENTS

vii

Chapter 1

THE HOUSE OF SCIENCE

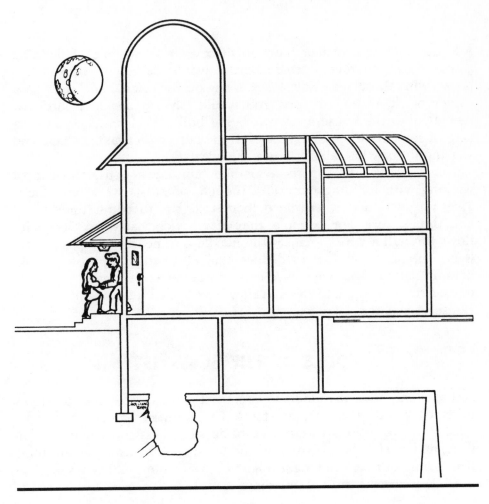

The House of Science is being built with scientific facts and tools. Using the scientific method as a blueprint, new rooms are added to the house as we come to better understand the world around us. As you will discover, the House of Science has changed and will continue to change our lives.

I f you and some friends were to discover a scrap pile of building materials, you might decide that it would be fun to construct a house with those materials. As you uncovered each item, such as a cinder block, rafter, or door, you would have to decide where that item fit into the house as it was being built. With patience and the help of some tools, you might soon find your project taking shape and starting to look like a real house (Figure 1.1)!

The work scientists do is a lot like building a house. Instead of working with building materials, though, scientists work with facts. They uncover facts, then piece them together to better understand how things work. For example, they may notice that when electricity flows through a wire, it causes the needle of a nearby compass to deflect. They use this fact to deduce that electricity is partially magnetic in nature. Thus they have come to better understand the nature of electricity. You could say that they have added a small piece to the House of Science.

TOOLS OF THE SCIENTIST

Just as a carpenter uses tools to construct a house, scientists use tools to uncover and piece together facts. The scientist's tools are, for the most part, extensions of our five basic senses—sight, sound, touch, taste, and smell. The microscope and telescope, for example, are tools that extend our sense of sight. Likewise, the water-quality tester (Figure 1.2) is an extension of our sense of taste.

FIGURE 1.1 The House of Science

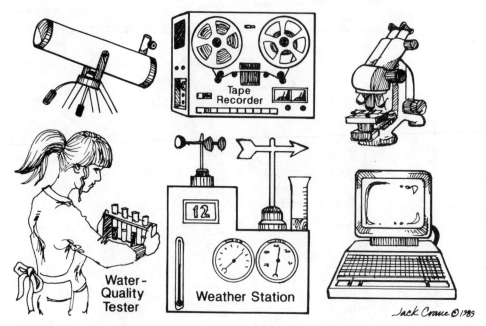

Tape
Recorder

Water-
Quality
Tester

Weather Station

Jack Crane © 1989

FIGURE 1.2   Some Tools of the Scientist

1. Some undersea creatures, such as whales and porpoises, communicate with one another using a complex system of sounds. To better understand their "language," which instrument in Figure 1.2 could you lower over the side of a ship to give you "ears" in the ocean?

Although the instruments shown in Figure 1.2 are all useful, it is the human mind that is the scientist's most important tool in constructing the House of Science. A good scientist, therefore, is someone who strives to keep a sharp and creative mind, by reading, for example. Indeed, the great British mathematician and philosopher Bertrand Russell once said that he used to read detective novels to keep his mind sharp.

ACTIVITY

Read a detective story and analyze how the detective put the facts together to solve the case. What clues were there to indicate the type of weapon used, for example. Or, what did the clothing each person wore reveal about that person? (The Sherlock Holmes stories by Sir Arthur Conan Doyle are good for sharpening your detective skills.)

THE SCIENTIFIC METHOD

If you were indeed to build a house, you would find it helpful to have a set of directions to guide you in its construction. These directions, called a blueprint, would tell you where the foundation is to be laid, for example, as well as where the walls and support beams should be placed. Without such a blueprint, you might soon find yourself in a house that wasn't very sturdy.

When building the House of Science, we also rely on a blueprint to guide us in its construction. This blueprint, called the *scientific method*, is a way of piecing together facts so as to understand the world around us. Using the scientific method, we accept only those explanations that we can verify over and over again by experiments.

To better understand the workings of the scientific method, let's look at how it was used by the great scientist Galileo (1564-1642).

During Galileo's lifetime, people accepted the wisdom of ancient philosophers, such as the Greek philosopher Aristotle. For example, Aristotle stated that, based on logic, the heavier an object is, the faster it will fall to the ground. A cannonball, for example, could be expected to fall many times faster than, say, a door key, because it is so much heavier. This logic seemed reasonable and was therefore accepted by people of the time. Galileo, however, refused to accept this and instead devised an experiment to see if Aristotle was indeed correct. As the story goes, Galileo dropped objects of different weights off the Leaning Tower of Pisa (Figure 1.3). He found that, neglecting air resistance, the objects all fell to the ground at the same time! This, of course, was directly opposite to what Aristotle had stated would happen.

2. (a) According to Aristotle, which object should fall to the ground first, a cannonball or a key?

(b) What would Galileo have found when he dropped both of these objects at the same time?

In the true form of the scientific method, Galileo's findings can be duplicated by further experiments. We can duplicate his findings by

Aristotle: "Logic shows that the heavier an object is, the faster it will fall to the ground."

Galileo: "Experiment shows that all objects will fall to the ground with the same speed!"

FIGURE 1.3   Aristotle's Method of Logic versus Galileo's Method of Experiment

dropping objects of different weights and noting whether they land at the same time. Neglecting air resistance (which tends to slow the fall of lighter objects more than heavier ones), we will find that all objects *do* reach the ground at the same time! Thus, we can reproduce Galileo's findings.

This ability to reproduce others' experiments is crucial to the scientific method. In some ways, it is like testing one of the rafters in a house to make sure that it is solid and has been installed properly.

3. In a vacuum, where there is no air (and hence no air resistance), which would fall to the ground faster, a feather or a brick?

The scientific method can be used in many aspects of our lives. In fact, many of us use this method without realizing it. If we encounter a strange dog, for example, we may extend a cautious hand toward it. This is an experiment we are conducting. If the dog seems receptive to this gesture, we may form a theory that the dog is friendly. If, however, the dog barks and growls, we may conclude that the dog is not friendly and decide to leave it alone.

4. Many professions require people to think scientifically. Doctors, for example, use the scientific method when they uncover symptoms, then piece them together to arrive at a diagnosis. In similar fashion, mechanics make use of this method when you take a car to them for repair. How do these mechanics use the scientific method in repairing your car?

ACTIVITY

Many companies make claims about a particular product they sell. Pick one of these products and devise an experiment that will prove or disprove a particular claim the company is making about it. For example, you might compare the strength of different cleaning products, or see if some batteries last longer than others.

SCIENCE IN OUR LIVES

As building proceeds on the House of Science, we find that it changes our daily lives. For example, many scientific discoveries are incorporated into our modern technology, so that today we have automobiles, computers, and countless other inventions that make our lives easier. Thanks to modern science, the world's best music can be heard by all, and long hours of tedious work can be saved by machines that do the work for us. Also thanks to science, modern medicines treat diseases that once had no cure. On the dark side, however, we also have such things as nuclear weapons and increased pollution as a result of modern science. As you will see, how we use science in the future is one of the challenges that will face all of us.

SUMMARY

The House of Science is being built as scientists uncover facts and piece them together into a framework of understanding. To assist them in this task, they use tools such as the microscope and telescope—much like a carpenter uses tools such as a hammer and saw. And to guide them in building the house, scientists use a blueprint known as the scientific method. Using this method, scientists accept as fact only that which they can prove true time after time by experiment.

As building proceeds on the House of Science, we find that it changes our daily lives. Discoveries in science have given us modern technology, for example, with both its good and evil aspects.

5. Sometimes parts of the House of Science have had to be torn down because the pieces (facts) did not fit together as we had originally thought. For example, at one time it was believed that the entire solar system (and universe) revolved around our own earth. Complicated theories were worked out to account for the unusual paths planets took as they supposedly circled us. Later it was discovered (by Galileo) that the planets' paths were not unusual at all, if instead of orbiting the earth, they orbit another object in our solar system. What object?

ANSWERS

1. The tape recorder

2. (a) A cannonball
   (b) They both reached the ground at the same time.

3. They would both fall to the ground at the same time.

4. They make a note of the car's "symptoms," then piece these symptoms together to come up with a "diagnosis" of how to fix it.

5. The sun

8

Chapter 2

MATHEMATICS

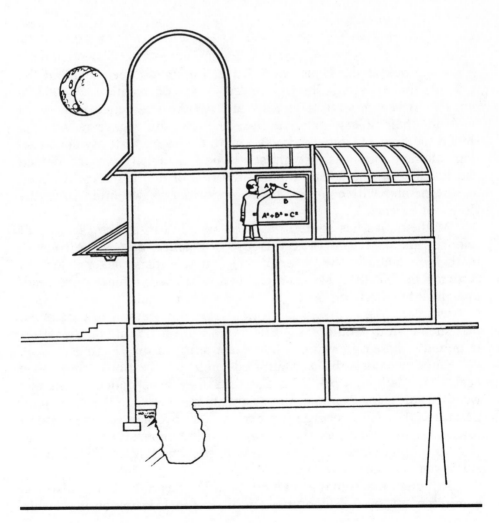

Mathematics plays a key role in building the House of Science. Sometimes called the "servant of science," mathematics allows scientists to quantify and work with the data they collect. Mathematics also has a "mind of its own" in that it sometimes predicts things before they occur and other times defies our common sense.

As we look at the House of Science, we should keep in mind the role that mathematics has played in its construction. For example, the surveyors who originally laid out the site relied heavily on math for their precise calculations. Likewise, the engineers who designed the house's framework used math to ensure a sturdy construction. Mathematics was also present when electricity was harnessed and installed in the house. Indeed, the role of mathematics has been important and will continue to be important as we build onto the House of Science.

Mathematics has sometimes been referred to as the "servant of science." As a "servant" in the House of Science, one of the roles of math is to quantify the observations that scientists make. Scientists studying the weather, for example, can take exact temperature readings, instead of relying on vague terms such as "mild," "quite warm," or "very cold." Of course, such terms vary from person to person, so that such a study would have little value were it not for mathematics to quantify these observations into exact temperature readings.

Once scientists have quantified their observations, they once again use math to work with the data they have collected. In our weather study, for example, the scientists may average the daily temperatures they have recorded to arrive at an average yearly temperature. They may then plot the average yearly temperatures for several years to observe changes in our climate over time. At each step in the study, mathematics plays a crucial role.

Although mathematics can be thought of as a servant, it also has a mind of its own. Sometimes, math is the forerunner in certain scientific discoveries. The discovery of radio waves, for example, was predicted mathematically by the British scientist James Clerk Maxwell in 1864. This was well before these waves were actually discovered (by Heinrich Hertz) in 1887!

Mathematics also has a mind of its own in that it sometimes refutes what we might take as common sense. It can be shown mathematically, for example, that if 23 people are in a room, there is better than a 50 percent chance that at least two of them will have the same birthday! Indeed, with 30 people in the room, there is about a 70 percent chance of this happening, and with 50 people, the probability goes up to 97 percent. Although this may defy your common sense, mathematics would be right and your common sense would be wrong!

ACTIVITY

In a room of people, have each person call off his or her birthday. Write these birthdays down and note if any of them match. How many people are in the room?

In this chapter we will look at some of the branches of mathematics. Keep in mind when you study math that it is in some ways like learning a foreign language. That is, it must be studied before you can understand it. As you read a math book, you may find that you can cover only a few pages (or less) in an hour. You may have to study the material closely while you build a foundation in your mind of what it all means. If you *do* take the time to build this foundation, you will find that it serves you well in better understanding all of the sciences.

ARITHMETIC

Arithmetic is the branch of mathematics that we are most familiar with. Its four major operations—addition, subtraction, multiplication, and division—are used in many aspects of our everyday lives. Although much of arithmetic seems "cut and dried," it sometimes requires you to think things through, also. Consider the following problems:

1. Twenty years used to be referred to as a "score" of years. In 1863, when Abraham Lincoln began his Gettysburg Address with "Four score and seven years ago," what year was he referring to?

2. A car's odometer reads 57,843 miles (93,085 km) when the gas tank is filled. When it is next filled, the odometer reads 58,170 miles (93,611 km) and it takes 8.6 gallons (33 liters) of gas. How many miles per gallon (km/liter) did the car get out of that last tankful of gas? (*Hint*: This problem requires first subtraction, then division.)

Figure 2.1 shows the number line. This line may be thought of as "home" for the numbers we work with. Each number has its place somewhere on this line, be it a whole number, a fraction, or—to the left of zero—a negative number. This line extends to infinity (forever) in either direction; there is no such thing as the largest number possible, for example. The number one could always be added to such a number, or it could be multiplied by two, three, or any number. Also, there are an infinite number of fractions between any two points on the line. Some of these fractions are shown between 0 and 1 on the number line. But even between the closer points of ¾ and 1, we could still fit an infinite number of fractions, such as ⁴⁄₅, ⁵⁄₆, ⁶⁄₇, ⁷⁄₈, and so on.

3. The number line can be used to add and subtract numbers. When adding 3 and 2, for example, we simply put our finger on the number 2 and "walk" 3 places to the right to the answer 5. Likewise, when subtracting 3 from 2, we simply walk 3 places to the left, giving us the answer of −1 (read "negative 1"). What is 2 minus 6?

The early Chinese mathematicians wrote positive numbers in black and negatives in red. This was the forerunner of the expression "in the red." If we say that a business is "in the red," we mean that it is in debt.

4. If you have $35 in the bank and write a check for $47, you have "bounced" the check. How much in debt would you be?

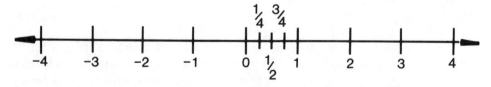

FIGURE 2.1   The Number Line

When dealing with fractions such as $1/3$, $1/4$, $2/7$, $11/16$, and so on, it is sometimes easier to convert them to decimals. This is done simply by dividing the denominator (bottom number) into the numerator (top number):

$1/3 = 1$ divided by $3 = .33\overline{3}$ (the 3 keeps repeating)
$1/4 = .25$
$2/7 = .29$ (carried to two decimal places)
$2/7 = .286$ (carried to three decimal places)

5. A batter who gets 6 hits in 15 at bats would have a batting average of $6/15$. What is this average in decimal form?

ALGEBRA

In many ways, algebra is a lot like arithmetic. But, instead of working entirely with numbers, in algebra we substitute letters for the quantities we don't know. We may write the following equation, for example:

$2 + x = 7$

In this equation, we can readily see that x stands for 5. In more complicated equations, however, we may not immediately know what the unknown (x) is. We would have to solve the equation to figure it out.

To better understand algebra, let's look at a practical problem that involves its use:

> *How many cuts would you have to make in a board to get two pieces of wood? To get three pieces? Four pieces? Five pieces?*

You can see from Figure 2.2 that the number of cuts you will have to make in the board is always one less than the number of pieces you end up with. Instead of stating this case by case, as was done in Figure 2.2, we could simply state that:

The number of pieces = the number of cuts made + 1

or, using abbreviations:

$P = C + 1$

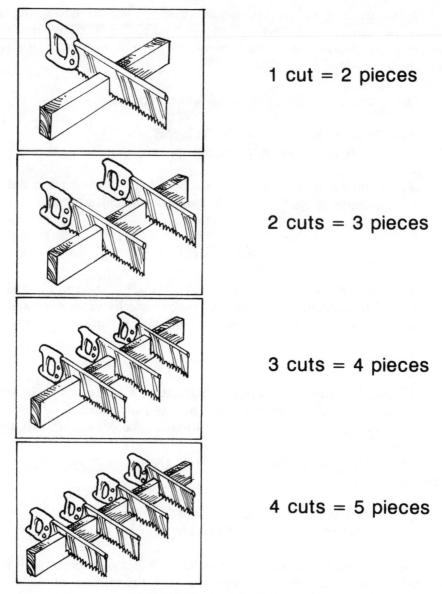

1 cut = 2 pieces

2 cuts = 3 pieces

3 cuts = 4 pieces

4 cuts = 5 pieces

FIGURE 2.2   Cutting a Board

We have just written an algebraic formula. With this formula, we can now solve a problem for any value of C or P. Let's try it:

If you need 24 pieces of wood, how many cuts will you have to make in the board?

We could do this problem in our heads, but let's use the formula instead:

$$P = C + 1$$

Substituting the P value that we know (24):

$$24 = C + 1$$

Now, since we are solving for C—the number of cuts that must be made—we must get this unknown quantity all alone on one side of the equation. We are not interested in what $C + 1$ is, after all, but rather in what C itself is. So we subtract one from both sides of the equation to come up with the answer:

$$24 - 1 = C + 1 - 1$$
$$23 = C$$
$$C = 23 \text{ cuts needed to get 24 pieces of wood}$$

6. If the number of pieces needed is 35, how many cuts will have to be made? Show your work by first writing down the original formula ($P = C + 1$), then "plugging in" the value given (35). Then solve for the unknown value (C).

Now let's say that the board we are cutting has knots in it so that only half of the pieces we cut are usable. We must modify our equation to account for the fact that we can use only half of the pieces:

$$P = (C + 1) \times \tfrac{1}{2}$$

The parentheses indicate that the entire quantity $C + 1$ is multiplied by $\tfrac{1}{2}$. Multiplying by $\tfrac{1}{2}$ is the same as dividing by 2, so our formula can be changed to this:

$$P = \frac{C + 1}{2}$$

Now let's say that 12 cuts are made in the board. How many *usable* pieces of wood will we get? Again, copy the formula, then plug in the given value.

$$P = \frac{C + 1}{2}$$

$$P = \frac{12 + 1}{2}$$

$$P = \frac{13}{2}$$

$$P = 6\frac{1}{2}$$

In this case, the answer $6\frac{1}{2}$ does not make good sense because a piece is either usable or not; it cannot be half-usable. Using common sense, we can reason that the remaining piece will either have a knot in it or it won't, so the answer is that 12 cuts will produce either 6 or 7 usable pieces.

7. If 17 cuts are made in the board, how many usable pieces will result? Show your work.

In the preceding examples with cutting a board, you used algebra to solve the problems. Although you were probably able to reason the answers out in your head, the equation served as a helpful tool. It would be good for you to learn how to use an equation, because you *will* need this knowledge to solve more complex problems.

In looking at problems you may encounter day to day, you can again use algebra as a tool to help you.

In the supermarket, a box of cereal costs $1.92 for a 12-ounce (340 g) box or $3.40 for a 20-ounce (570 g) box (Figure 2.3). Ounce for ounce, which is the better buy?

To find the cost per ounce of the small box of cereal, you can set up the following equation:

$$\frac{\$1.92}{12 \text{ oz.}} = \frac{\$x}{1 \text{ oz.}}$$

You can read this equation as $1.92 is to 12 ounces as x (we call the unknown x) is to 1 ounce. Now you can cross-multiply, as shown:

$$\frac{\$1.92}{12 \text{ oz.}} \diagdown \frac{\$x}{1 \text{ oz.}}$$

$1.92 \times 1 = 12 \times x$ *Note*: $12 \times x$ can simply be written $12x$.

$1.92 = 12x$

FIGURE 2.3 Two Boxes of Cereal

Now divide both sides of the equation by 12 so that the right side of the equation simply becomes $1x$ (written just as x):

$$\frac{1.92}{12} = \frac{12x}{12}$$
$$.16 = x$$
$$x = 16 \text{ cents per ounce}$$

(The cereal sells for 16 cents per ounce in the 12-ounce box.)

8. What is the price per ounce of the 20-ounce box of cereal that sells for $3.40? Create the equation by thinking $3.40 is to 20 ounces as x is to 1 ounce.

9. Ounce for ounce, which of the two boxes of cereal is the better buy, the 12-ounce (340 g) box or the 20-ounce (570 g) box?

The cross-multiplication method shown here is useful in many ways. For example, if a car trip of 280 miles (450 km) takes 7 gallons (27 liters) of gas, how much gas will be needed for a trip of 540 miles (850 km)? Set up the equation as follows:

$$\frac{7 \text{ gal.}}{280 \text{ mi.}} = \frac{x \text{ gal.}}{520 \text{ mi.}}$$
$$\left(\text{In metric: } \frac{27 \text{ l.}}{450 \text{ km}} = \frac{x \text{ l.}}{850 \text{ km}} \right)$$

17

Again, read the above equation as 7 gallons (27 liters) is to 280 miles (450 km) as x gallons (liters) is to 520 miles (850 km).

10. Cross multiply the equation, then find the number of gallons (liters) required by solving for x.

11. A certain recipe for four servings calls for 1 cup of sugar, but you are cooking for five. How much sugar should you use? Set up an equation using x as your unknown, then cross multiply and solve this equation.

You have been solving problems in algebra. You formulated equations, then solved them. This is what algebra is all about.

GEOMETRY

Geometric shapes can be seen all around us (Figure 2.4). Geometry is the study of these shapes.

12. Which object in Figure 2.4 is a circular disk? Which is a sphere?

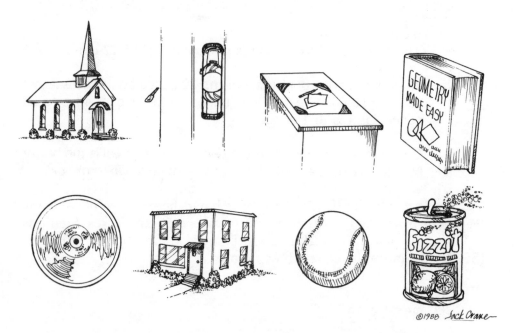

©1988 Jack Crane

FIGURE 2.4 Geometric Shapes in Our Environment

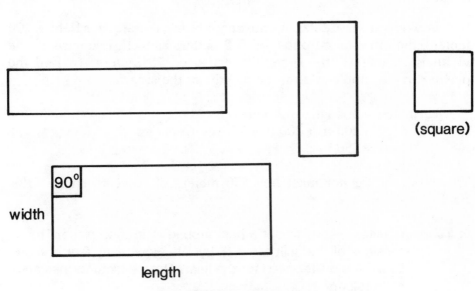

FIGURE 2.5   Rectangles

One of the simplest and yet most useful of geometrical shapes is the rectangle. As you can see in Figure 2.5, all rectangles have four sides. These sides in turn are perpendicular to each other. You may also notice that the opposite sides of rectangles are equal in length and are parallel.

To find the distance around a rectangle (its *perimeter*), simply add up the lengths of its four sides:

> *A farmer wants to put a fence around a rectangular field that is 200 meters long and 75 meters wide. How many meters of fence will be needed?*

To do this problem, first draw a picture of the field (Figure 2.6).

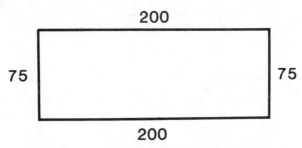

FIGURE 2.6   A Field 200 Meters by 75 Meters

19

Looking at Figure 2.6, if one length of the rectangular field is 200 meters, the other length must be 200 meters as well. Likewise, if one width is 75 meters, the other must also be 75 meters. To find the perimeter, we simply add up the lengths of the sides:

Perimeter = 2 lengths + 2 widths
P = 200 m + 200 m + 75 m + 75 m
P = 550 m

Therefore, the farmer must buy 550 meters of fence to enclose this field.

13. A quilter wishes to put a hem around a finished piece. The dimensions of the quilt are 7 ft. by 5 ft. How many feet of material will the quilter need for the hem? Draw a picture, then solve the problem.

14. A square is a type of rectangle in which all of the sides are of equal length. A baseball diamond is actually a square in which the distance between bases is 90 feet (27.44 m). What is the perimeter around the bases of a baseball diamond?

The area of a rectangle is the amount of space bounded by the sides of that rectangle. We define the *area* of a rectangle as the length times the width. Let's see if this definition holds true for the rectangles shown in Figure 2.7.

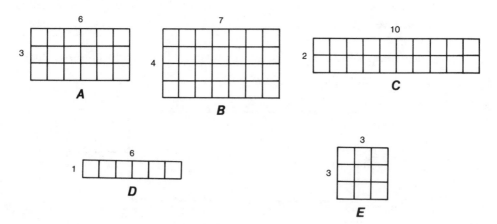

FIGURE 2.7 Areas of Rectangles

20

15. Rectangle A has an area of 18 square centimeters. This was figured simply by counting up the squares. Count up how many square centimeters are present in rectangles B, C, D, and E.

16. Multiply the length times the width for each of the rectangles in Figure 2.7. How do these answers compare with the answers you got by counting individual squares?

17. The area (*A*) of a rectangle is equal to the length (*L* times the width (*W*). Write this equation in mathematical form.

18. A map covers an area 19 kilometers (30 miles) by 12 km (20 miles). How many square km are shown on the map?

Like the rectangle, the circle is a common geometrical shape. Figure 2.8 shows the various parts of a circle.

Figure 2.8 shows that the *diameter* is any line segment that goes through the center point of the circle. The *circumference,* in turn, is the distance around the circle. You can think of it as the perimeter of the circle. Finally, the *radius* is one-half the diameter. Like the diameter, it also includes the center point of the circle.

Now pretend that you can take the diameter out of a circle and bend it around the circumference (Figure 2.9). How many diameters would be needed to go around the circle?

Notice from Figure 2.9 that it takes a little more than three "bent" diameters to go around the circle. You can try this with any size circle you like. The number of diameters required to go around the circumference will always be a little more than three. More precisely, mathematicians have figured that it would take about ($\sim$) 3.14 diameters. This number, ~ 3.14, is called *pi.* The symbol for it is π.

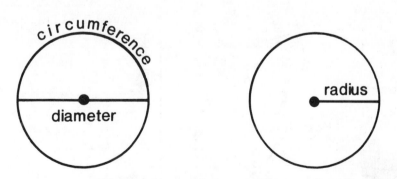

FIGURE 2.8   The Parts of a Circle

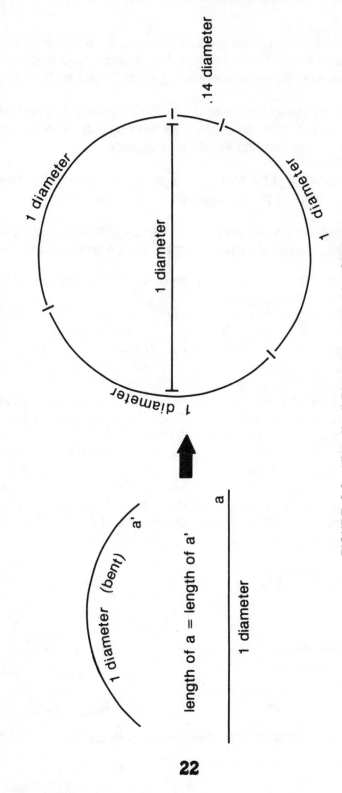

FIGURE 2.9 "Bending" Diameters around a Circle

ACTIVITY

Fit a piece of string around the circumference of a circular object, then straighten out the string and measure it. Its length should be pi times the diameter of that object.

In summary, it takes ~3.14 diameters to go around the circumference of a circle. In equation form, we could say that:

Circumference = $\pi \times$ diameter

or, in mathematical form

$C = \pi d$

19. If you assume that Earth is a sphere with a diameter of 8,000 miles (13,000 km), what is the distance around the equator? That is, what is the circumference of our planet? To solve this problem, draw a picture of a circle and label the diameter 8,000 miles (13,000 km). Then, referring to the equation given earlier, figure out the circumference.

To figure out the area (A) of a circle, use the following formula:

$A = \pi r^2$

where r^2 represents the radius times itself.

The formula $A = \pi r^2$ is not an easy one to come up with on your own, yet this formula is a useful tool to have nevertheless. Just as we use appliances that others have invented, we can use formulas that others have derived for us.

20. If an oil spill spreads over an area of water in a circular slick five miles (8 km) in radius, find the area covered by the slick. Draw a picture, then use the formula for area (A).

So far, we have been dealing with perimeters and areas. Geometry is useful in figuring volumes also. For example, to figure the volume of a rectangular box, simply take the length × width × height, as shown in Figure 2.10.

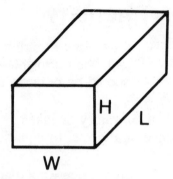

Volume (V) = L x W x H
or simply V = L W H

FIGURE 2.10   Volume of a Rectangular Box

21. A rectangular aquarium three feet (.9 m) long, two feet (.6 m) wide, and two feet (.6 m) high will hold how many cubic feet (cu. meters) of water?

22. If a cubic foot of water weighs 62.5 pounds (a cubic meter of water weighs 1,000 kg), how many pounds (kg) of water will the aquarium hold when it is completely filled?

In addition to finding the volumes of rectangular shaped objects, we can also find the volumes of other geometric objects. For example, the volume of a soup-can-shaped cylinder can be found by using the formula given in Figure 2.11.

23. If a cylinder-shaped grain silo has a 10-meter radius and a height of 50 meters, how many cubic meters of grain will it hold?

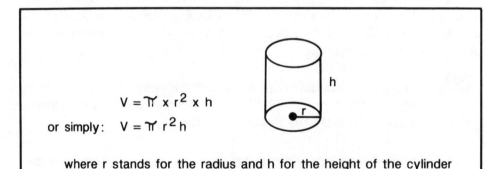

$V = \pi \times r^2 \times h$

or simply: $V = \pi r^2 h$

where r stands for the radius and h for the height of the cylinder

FIGURE 2.11   Figuring Volume for a Soup-Can-Shaped Cylinder

The problems you have been doing here should give you a good introduction to geometry. An understanding of geometry is very useful; it can be helpful in many walks of life. Engineers, architects, and artists, in particular, have many uses for this branch of mathematics.

24. Some artists like to "play games" with the principles of geometry. Why do you think the sketch in Figure 2.12 is called "Impossible Object"?

ANALYTIC GEOMETRY

Analytic geometry has been described as a "marriage" between algebra and geometry. Very simply, this branch of mathematics takes the equations from algebra and graphs them so that they can be seen as geometric shapes. You then not only have the equation, but a picture of it as well!

Let's look at a simple algebraic formula and see how it would be graphed. Let's say that on a certain test you are studying for, you will get four questions right for every hour you study. You could express this using algebra by the formula:

$Q = 4H$

where Q stands for the number of questions answered correctly and H for the hours studied.

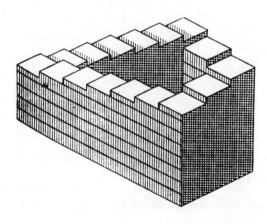

FIGURE 2.12 "Impossible Object" (© 1988 Cordon Art—Baarn—Holland)

TABLE 2.1
H and *Q* Values

H	Q
0	0
1	4
2	8
3	12
4	16
5	20

Table 2.1 shows various Q values for given values of H. Looking at this chart, we can see that if zero hours of studying are done ($H = 0$), zero questions will be answered correctly ($Q = 0$). Likewise, if one hour of studying is done ($H = 1$), four questions will be answered correctly ($Q = 4$). For two hours studying, eight questions will be answered correctly, and so on.

Figure 2.13 is a graph of the H and Q values given in Table 2.1. As seen on this figure, for the first value ($H = 0$, $Q = 0$), we simply put a point on the graph where the two lines meet. For the next point (1,4), we count over to the 1 on the hours studied line, then count up to the 4 on the line for questions answered correctly. We then plot the point.

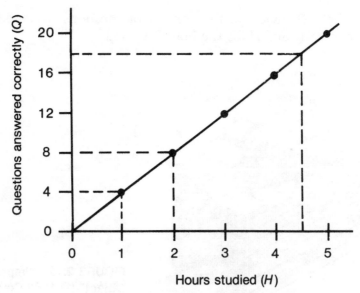

FIGURE 2.13   Graph of Equation $Q = 4H$

For the point (2,8), we count over to the 2 on the *H* line, then up to the 8 on the *Q* line. This has been done for all six values from the *H*-*Q* chart. Make sure you understand how these points were plotted.

Notice on Figure 2.13 that after our points have been plotted, we can then draw a straight line that runs through all of them. From this line we can then figure out intermediate values of *H* and *Q*. We can see, for example, that if four-and-one-half hours are spent studying, 18 questions will be answered correctly (see dotted line on graph). We could, of course, figure this out using the equation also.

25. Use the graph to figure out how many questions will be answered correctly if two-and-one-quarter hours of studying are done.

26. Check your answer to question 25 using the equation $Q = 4H$. (*Note*: 2¼ hours = 2.25)

Now let's graph an equation that is a little different. Take the equation:

$$y = x^2 \qquad \text{Note: } x^2 \text{ means } x \text{ times } x.$$

This equation approximates the path that a comet might take around the sun, for example. To graph this equation, let's again set up a chart of values for *x* and *y* (Table 2.2).

Notice from Table 2.2 that when $x = 0$, $y = 0$. Likewise, when $x = 1$, $y = 1$, because x^2 (1 × 1) still equals one. For the negative numbers on this chart, we get a positive *y* value when we square them. This is a fundamental rule of arithmetic—a negative number times a negative number (even that same number) always yields a positive number.

TABLE 2.2
x and y Values

x	y
0	0
1	1
−1	1
2	4
−2	4
3	9
−3	9

27

27. (a) What is -3×-3?

(b) What is -3×-5?

(c) What is -6^2

Now we are ready to graph the equation (Figure 2.14). The graph looks a little different because we must account for negative values of x as well.

Figure 2.14 shows that all seven points from Table 2.2 have been plotted. Arrows were put on the graph to show that it extends outward in both directions. Also, we drew a curve between the points, instead of connecting them with straight lines. If this were indeed the path of a comet, we would expect the path to be curved. Figuring the equation for more points would, of course, give us the curved path also.

28. From the graph in Figure 2.14, what is the approximate value for y when x is 2.5?

29. Using the equation $y = x^2$, what is the value for y when $x = -2.5$?

Like algebra and geometry, analytic geometry is a useful tool to have. An understanding of graphs, particularly, will be very helpful to you.

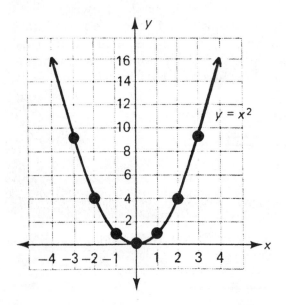

FIGURE 2.14 **Graph of Equation $y = x^2$**

PROBABILITY AND STATISTICS

This branch of mathematics got its start when gamblers wanted to figure out their odds in games of chance such as cards and dice. By knowing the odds, they hoped to gain an advantage over their opponents (Figure 2.15).

"Knowing the odds" has since grown into a science with many applications. Whenever you say, "I wonder what the chances of this happening are?", you are engaging in an exercise in probability.

To figure out probabilities, statistics must be used. *Statistics* are simply facts that someone has gathered, then put into categories so that we can use them. When we flip a coin, for example, we form two categories for the possible outcome. These categories are "heads" and "tails." From the statistics we gather as we flip the coin, we can then

FIGURE 2.15   W.C. Fields, the Poker Expert, in *My Little Chickadee* (Copyright © by Universal Pictures, a Division of Universal City Studios, Inc. Courtesy of MCA Publishing Rights, a Division of MCA Inc.)

determine probabilities. In this case, if the coin is flipped enough times, the heads and tails categories will be close to equal, and we can determine that both heads and tails have a probability of coming up about 50 percent of the time.

Now let's look at dice. When a die is thrown, there are six possibilities of how it can land. These possibilities are, of course, 1, 2, 3, 4, 5, or 6. When throwing two dice, you may have noticed that certain numbers come up more frequently than others. The number 7, for example, comes up more frequently than, say the number 12. To see why this is true, let's make a chart of all the possible outcomes of throwing two dice (Figure 2.16).

You can see on Figure 2.16 that there are 36 different possible combinations for throwing two dice. For example, looking at the chart, you will notice that if a six lands on one die and a one on the other, the outcome will be seven. Likewise, a one on the one die and a six on the other also produces a seven.

30. In addition to a six and one and a one and six, what other four ways can the dice come up so that the total will be seven?

31. What six possible numbers can you get when you throw doubles?

outcome of the one die

		1	2	3	4	5	6
	1	2	3	4	5	6	7
	2	3	4	5	6	7	8
outcome of the other die	3	4	5	6	7	8	9
	4	5	6	7	8	9	10
	5	6	7	8	9	10	11
	6	7	8	9	10	11	12

FIGURE 2.16   The 36 Possible Outcomes of Throwing 2 Dice

By looking at Figure 2.16, we can tally how often a certain number is likely to come up. The number two, for example, has only 1 chance in 36 of coming up on any given roll. The number three, on the other hand, has 2 chances in 36 of coming up (a three and a one, or a one and a three).

32. List the numbers 2 through 12 on a sheet of paper. Then, beside each number, give the number of times it is likely to come up in 36 rolls of the dice.

The numbers you just compiled could be presented in a lot of ways. Figure 2.17 shows two common ways statisticians use to present data they have tabulated.

33. Looking at the bar graph in Figure 2.17, we see that the number four is likely to come up an average of 3 times in 36 rolls. What other number has this same likelihood of coming up?

34. Referring to the pie diagram, what are the chances that either a six, seven, or eight will come up on any given roll of the dice?

Many people who play games of chance have the mistaken notion that if a certain number does not come up for some time, it is "due." Nothing could be further from the truth. If "snake eyes" (two ones),

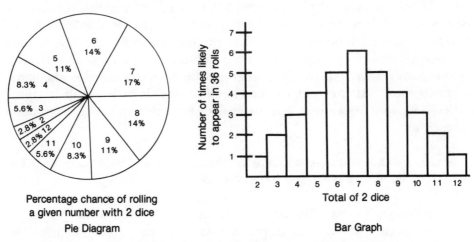

Percentage chance of rolling
a given number with 2 dice

Pie Diagram

Bar Graph

FIGURE 2.17   Pie Diagram and Bar Graph of Data from Figure 2.16

for example, does not come up in 50 rolls of the dice, the likelihood of it coming up still remains at 1 chance in 36. The dice, after all, do not possess a memory. To think that they do is simply foolish.

35. A flipped (honest) coin has come up tails five times in a row. Since you have been betting on heads, you may think that heads is now "due." What are the chances of heads coming up on the next throw?

Statisticians are often concerned with the average value for data they have collected. If a test is taken by five students, for example, out of 10 possible correct answers; the results may be:

3, 7, 7, 8, 10

To find the average for this data, simply add the numbers and divide by the number of students taking the test.

36. What is the average test score for the five students taking the test?

37. If you could throw out the low score of three, what would be the average for the remaining four test scores?

Statisticians sometimes present data in a misleading fashion. They are not actually lying about the data, they are simply casting it in a light favorable to their cause (Figure 2.18).

There are two graphs in Figure 2.18, each of which presents the same data. In the top graph, the percentage scale is shown only between 47 and 50 percent, whereas on the bottom one it is shown between 0 and 100 percent. The top graph makes it appear that Honest Joe is headed for real trouble, whereas the bottom one uses the same data to show that his support has been steady. Which graph is right? They both are! It is up to you to realize how the statistics have been presented.

38. About what percentage of the vote did Honest Joe lose between January and the end of June?

Although probability and statistics grew up out of games of chance, today they play an important role in many walks of life. The insurance industry, for example, bases its rates on the probabilities of

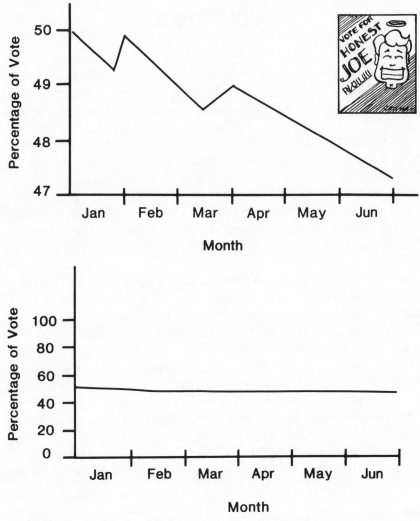

FIGURE 2.18 **Percentage of People Supporting a Candidate**

certain accidents or disasters happening. Statistics of what has happened in the past are used to calculate these probabilities.

ACTIVITY

Write an insurance policy for the House of Science. What possible disasters should you insure against? How will you determine your rates?

33

COMPUTERS

No discussion of mathematics would be complete without some mention of computers. Ranging in size from small pocket calculators to room-sized instruments, computers have been an immensely valuable tool to the mathematician. Tedious computations that might take hours on paper, for example, can now be done in seconds on the computer.

Contrary to what some people think, computers do not have minds of their own. They are programmed by people and will do exactly what they are told—nothing more and nothing less.

Figure 2.19 shows a flowchart for a computer program. This program is designed to compute the wages for individuals working at a company. If several hundred employees worked at this company, this computer program could save many hours of work figuring out their paychecks.

A *flowchart* simply shows the steps the computer must go through to come up with the answer. In looking at Figure 2.19, we can see that for any given employee, the following information must be fed in: name, number of hours worked, and hourly rate. The computer then asks if the employee worked over 40 hours or not. If he or she has not, the salary (base pay) is simply found by multiplying the hourly rate times the number of hours worked. The computer then prints out the employee's name and wage on his or her paycheck.

If the employee has worked more than 40 hours, the computer must first calculate the overtime pay, then the base pay. It then adds these together and prints out the employee's name and salary as before. Notice that the overtime pay is one-and-one-half times (time-and-one-half) the normal hourly rate.

39. To do this problem, you will have to "walk" through the computer program yourself. Employee Bob Price makes $6 per hour and works 45 hours one week. What will his weekly paycheck be when these statistics are fed in?

SUMMARY

Mathematics plays a crucial role in the House of Science. This "servant of science" allows researchers to quantify the data they collect and, in turn, work with this data. In addition to being a servant, however, mathematics also has a "mind of its own" in that it sometimes predicts

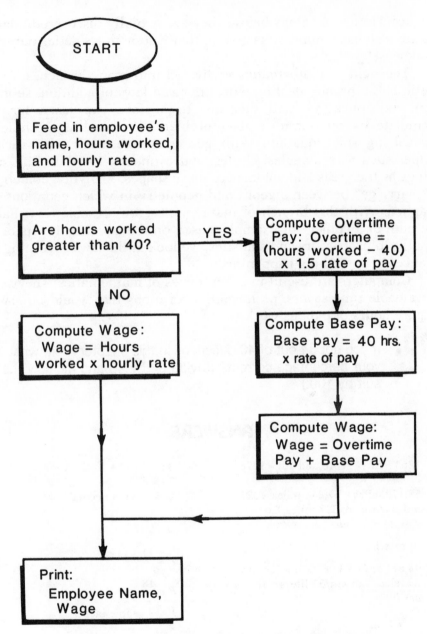

FIGURE 2.19   Flowchart for a Computer Program

the occurrence of things before they are actually discovered. Mathematics also has a mind of its own in that it sometimes defies our common sense!

The field of mathematics is divided into many branches. Arithmetic is the branch dealing with the basic laws of addition, subtraction, multiplication, and division. In algebra, we learn how to formulate an equation for a given problem. We then learn techniques for solving this equation. With geometry, we work with various shapes such as rectangles, circles, and cylinders. We then find such things as the areas and volumes of these shapes. Analytic geometry is a "marriage" between algebra and geometry in which equations are graphed. With this branch of mathematics, we are able to visualize the path of a comet, for example, based on the equation for its orbit. Finally, in probability and statistics, we look at the laws governing the chances of something happening.

Computers are useful in all branches of mathematics. They serve as valuable time savers, performing operations that would otherwise require hours of tedious work.

40. If you got 34 out of 40 questions in this chapter right, what was your percentage correct? (Divide 34 by 40 and multiply the result by 100.)

ANSWERS

1. 1863 – 87 (4 score and 10) = 1776

2. 58,170 miles – 57,843 miles = 327 miles traveled; 327 miles/8.6 gallons = *38 miles per gallon*

In metric:

93,611 km – 93,085 km = 526 km traveled; 526 km/33 liters = *16 km per liter*

3. 2 – 6 = –4

4. $12

5. .400

6. $P = C + 1$
$35 = C + 1$

$35 - 1 = C + 1 - 1$
$34 = C$
$C = 34$ cuts needed

7. $P = \dfrac{C + 1}{2}$

$P = \dfrac{17 + 1}{2}$

$P = \dfrac{18}{2}$

$P = 9$ usable pieces

8. $\dfrac{\$3.40}{20 \text{ oz.}} = \dfrac{x}{1 \text{ oz.}}$

$\$3.40 \times 1$ oz. $= 20$ oz. $\times x$
(cross-multiplication)
$3.40 = 20x$
$\dfrac{3.40}{20} = \dfrac{20x}{20}$

$.17 = x$

$x = 17$ cents per ounce

9. In this case, the smaller box would be the better buy (16¢/oz. vs. 17¢/oz. for the larger box).

10. $\dfrac{7}{280} = \dfrac{x}{520}$

$280x = 7 \times 520$
   (cross-multiplication)
$280x = 3640$
$\dfrac{280x}{280} = \dfrac{3640}{280}$
$x = 13$ gallons

In metric:

$\dfrac{27}{450} = \dfrac{x}{850}$

$450x = 27 \times 850$
   (cross-multiplication)
$450x = 22{,}950$
$\dfrac{450x}{450} = \dfrac{22{,}950}{450}$
$x = 51$ liters

11. $\dfrac{1 \text{ cup}}{4 \text{ servings}} = \dfrac{x \text{ cups}}{5 \text{ servings}}$

$4x = 5 \times 1$ (cross-multiplication)
$4x = 5$
$\dfrac{4x}{4} = \dfrac{5}{4}$
$x = \dfrac{5}{4}$ cups
or $x = 1\,{}^{1}/_{4}$ cups

12. The circular disk is the record; the sphere is the baseball.

13. $P = 3 + 3 + 2 + 2$
   $P = 10$ meters

14. $P = 90 + 90 + 90 + 90$
   $P = 360$ feet (109.8 m)

15. B has 28 sq. cm. C has 20 sq. cm. D has 6 sq. cm. E has 9 sq. cm.

16. They are exactly the same.

17. $A = L \times W$

18. $A = L \times W$
   $A = 19$ km $\times 12$ km
   $A = 228$ sq. km (600 sq. miles)

19. $C = \pi d$
   $C = \pi \times 8{,}000$ miles
   $C = 3.14 \times 8{,}000$ miles
   $C = 25{,}120$ miles

In metric:

$C = \pi d$
$C = \pi \times 13{,}000$ km
$C = 3.14 \times 13{,}000$
$C = 40{,}820$ km

20. $A = \pi r^2$
   $A = \pi \times 5$ miles $\times 5$ miles
   $A = \pi \times 25$ sq. mi.
   $A = 78.5$ sq. mi.

In metric:

$A = \pi r^2$
$A = \pi \times 8$ km $\times 8$ km
$A = \pi \times 64$ sq. km
$A = 201$ sq. km

21. $V = L \times W \times H$
   $V = 3 \times 2 \times 2$
   $V = 12$ cubic feet

In metric:

$V = L \times W \times H$
$V = .9 \times .6 \times .6$
$V = .324$ cu. meters

22. 12 cu. ft. $\times$ 62.5 lbs./cu. ft. = 750 lbs.

In metric:

.324 cu. meters $\times$ 1,000 kg/ cu. meter = 324 kg

23. $V = \pi r^2 h$
   $V = \pi \times 10$ meters $\times 10$ meters $\times 50$ meters
   $V = 15{,}700$ cu. meters

37

24. It would be impossible to build an object in which the steps continually rose (or fell, depending on how you look at it).

25. Nine questions will be answered correctly.

26. $Q = 4H$
$Q = 4 \times 2.25$
$Q = 9$ questions answered correctly

27. (a) 9
(b) 15
(c) 36

28. The value is a little over 6.

29. $y = x^2$
$y = (-2.5)^2 = -2.5 \times -2.5$
$y = 6.25$

30. 5,2
2,5
3,4
4,3

31. 2, 4, 6, 8, 10, 12

32. 2—once, 3—twice, 4—three times, 5—four times, 6—five times, 7—six times, 8—five times, 9—four times, 10—three times, 11—twice, 12—once

33. 10

34. 6 - 14%; 7 - 17%; 8 - 14%; 45% total chance

35. 50 percent chance (no more, no less)

36. $A = \dfrac{3 + 7 + 7 + 8 + 10}{5}$
$A = \dfrac{35}{5}$
$A = 7$

37. $A = \dfrac{7 + 7 + 8 + 10}{4}$
$A = \dfrac{32}{4}$
$A = 8$

38. About 3 percent (from 50 to almost 47 percent)

39. 5 hours overtime = 5 hrs. $\times$ \$6/hr. $\times$ 1 ½ = \$45
40 hours regular = \$6/hr. $\times$ 40 hrs. = \$240
regular (\$240) + overtime (\$45) = \$285
Bob Price's weekly pay will be \$285.

40. $\dfrac{34}{40}$ = 85 percent

Chapter 3

CHEMISTRY

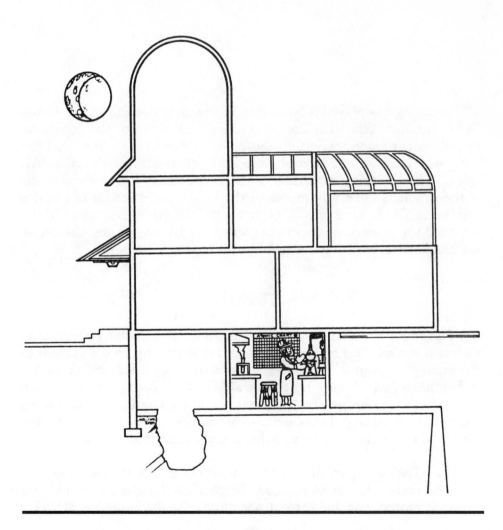

Chemistry provides many of the building materials for the House of Science, as well as the products used in the house. Chemistry has many applications: it is used to separate substances, to uncover clues in a police investigation, and to study the environment and how to improve it.

C hemistry has played a large role in the construction of the House of Science. When the concrete foundation was poured, for example, chemical reactions within the cement caused it to harden. Likewise, when the walls and roof were built, chemistry produced sturdy and durable building materials. Finally, chemistry was behind many of the products currently in use in the house. Items made of plastics and metal alloys, for example, were all developed by chemists. To understand the science of chemistry, let's look first at atoms, the building blocks of everything we see around us.

ATOMS

Use Figure 3.1 to visualize the atom as a miniature solar system, with a central nucleus and one or more orbiting electrons. In the nucleus of the atom are protons, which have a positive electrical charge of +1, and neutrons, which have no charge at all. The orbiting electrons each have a negative charge of −1, so that, for any given neutral atom, the number of electrons always equals the number of protons. Thus, the electrical charges in the atom balance each other out.

1. The hydrogen atom is the simplest of all the atoms. It has only one proton in its nucleus. To balance the charge of this lone proton, how many electrons must orbit the hydrogen atom?

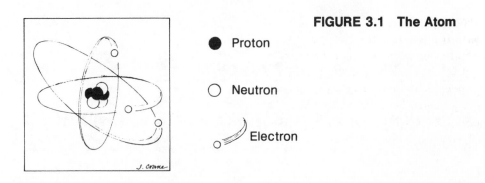

FIGURE 3.1 The Atom

● Proton

○ Neutron

Electron

J. Crane

2. A uranium atom has 92 protons and 146 neutrons in its nucleus. This makes uranium one of the larger atoms. How many electrons orbit the nucleus of this atom?

Atoms are incredibly small—so small, in fact, that two million of them lined up end to end would barely span the period at the end of this sentence. Although they are small, atoms account for everything we see and feel around us. The chairs we sit on are made of atoms, as are the clothes we wear and the air we breathe.

There are 109 different kinds of atoms, 92 of which can be found in nature. The remaining 17 have been created only in the laboratory. These 109 types of atoms make up the various elements, such as hydrogen, oxygen, iron, mercury, and gold.

3. What are the three main particles that make up an atom of any of the 109 elements?

The story of chemistry can be told with the electrons that circle the nucleus of the atom. These electrons do not circle in any random path. Rather, they circle in certain predictable orbits called *shells*. Like a parking garage filling from the bottom up, these electrons fill their shells in a very predictable way (Figure 3.2).

In Figure 3.2, the number of protons in the nucleus is shown in the center of each atom. Remember, for the atom to balance electrically, the number of electrons must equal the number of protons. This is because protons have a charge of + 1, while electrons have a charge of −1.

Looking at Figure 3.2, we see that the one electron circling the hydrogen atom is to be found in shell 1, the innermost shell. This is as we would expect. Likewise, the two electrons of the helium atom also occupy the first shell. Since this first shell can hold only two electrons, we say that it is *full*.

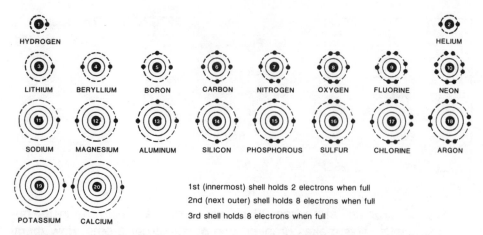

FIGURE 3.2 Electron Shells (© Compton's Encyclopedia, A Division of Encyclopedia Britannica)

An atom of lithium has three electrons: two of them in shell 1, making that shell full, and the third in shell 2, the next outer shell. Notice that for this element the inner shell does not show its two electrons. Instead, a complete circle is used to indicate that shell 1 is filled with electrons.

As you can see in Figure 3.2, the second shell continues filling with electrons for the elements beryllium through neon. Neon, with its ten electrons, completely fills both shells 1 and 2.

4. (a) Of the seven electrons in nitrogen, how many are in shell 1?

   (b) How many electrons occupy its second shell?

   (c) How many electrons does the second shell hold when it is full?

Sodium, with its 11 electrons, begins filling shell 3. This shell continues filling until it is full with 8 electrons. As you can see, this occurs with the element argon.

5. Silicon (#14) has how many electrons in each of its three electron shells?

6. Elements with the same number of electrons in their *outermost* shells often have many similar properties. For example, look at nitrogen and phosphorus, two similar elements used in fertilizers. How many electrons does each of these elements have in its outermost shell?

The remainder of the 109 elements also have shells occupied by electrons. Some of these shells may hold as many as 14 electrons. The important thing to remember is that these electrons are confined to specific shells. They are not free to orbit the nucleus of the atom in any random path. Indeed, the way these electron shells are filled, particularly the outer shell, determines many of the properties of an element.

7. Lithium is a highly reactive substance, as is the element sodium. How many electrons do each of these elements have in their outermost shells?

8. The outer shell of neon, a gas, is full with electrons, which makes this element very stable. From Figure 3.2, name two other elements whose outer electron shells are also filled (and as a result are also very stable).

Figure 3.3 shows the periodic table of elements. This table lists the 109 elements presently known. These elements are listed according to the number of protons present in the nucleus of each element. For example, element 53, iodine, has 53 protons in its nucleus.

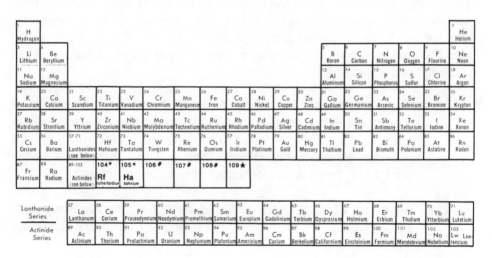

* name not yet official
\# name not yet proposed
★ synthesis not yet confirmed

FIGURE 3.3 The Periodic Table of Elements (Base Table: Compton's Encyclopedia, vol. 18, p. 207 © 1989 "Periodic Table")

The elements shown in Figure 3.3 are grouped according to how their electron shells are filled. Those elements to the extreme right side of the figure, for example, all have their outer shells completely filled with electrons. These are the elements helium, neon, argon, krypton, xenon, and rubidium. On the other hand, the elements to the extreme left of the chart have only one electron in their outermost shells.

Notice on Figure 3.3 that elements 3 and 4 (lithium and beryllium) appear by themselves to the left of the chart. They have been separated from the other six elements that store electrons in that second electron shell (B, C, N, O, F, and Ne). Likewise, elements 11 and 12 (sodium and magnesium) have also been separated from the other six elements that store electrons in the third electron shell. The reason for the separation is because these elements share similar properties with other elements farther down on the chart. As you just saw, Li and Na, for example (as well as hydrogen), have only one electron in their outermost shells. This makes them similar to the rest of the elements in that column (K, Rb, Cs, Fr).

9. Beryllium and magnesium are similar in that they have two electrons in their outermost shells. Name four other elements in that column that share this property.

After the element calcium (#20), the filling of the electron shells becomes more complicated. This is not to say that it is not orderly, however. It is like a parking garage that has a lot of nooks and crannies for parking spaces. And nature dictates the exact order in which these spaces are filled.

10. The Lanthanide Series and Actinide Series at the bottom of Figure 3.4 each represent large electron shells. Each new element in one of these series holds one more electron than the previous element. With which element is the Lanthanide Series completely filled with electrons? the Actinide Series?

11. Elements 21 to 30 and 39 to 48 each fill fairly large electron shells. How many electrons can be held in each of these shells?

As you can see in Figure 3.3, many of the symbols for elements have been taken from one or two letters in their English names. Oxygen, for example, is given the symbol O, and magnesium the symbol Mg. Some of the chemical symbols, however, are taken from the Latin words for the elements. Fe, for example, is short for the Latin word *ferrum,* which means iron. Likewise, Pb is short for the Latin word *plumbum,* which means lead.

12. In 1964, the Republican candidate for president of the United States passed out bumper stickers that read "AuH_2O." What was the candidate's name?

ACTIVITY

Make a set of flash cards to help you learn the various elements. Put the symbol for the element on the front of each card and the name of the element on the back.

COMPOUNDS

The elements shown in Figure 3.3 may be thought of as the cast of players in a show. Some of these elements, such as hydrogen, oxygen, and carbon, appear very frequently, while others have small and, to date, insignificant parts. Some elements, such as uranium and radon, play unique roles in that they are radioactive, and therefore break down into other elements with time.

Chemistry would be a simple matter if everything were made of pure elements. Very few things in nature appear in their elemental form, however. Instead, elements combine to form compounds. Why this occurs can be shown with the simple compound sodium chloride—table salt (Figure 3.4).

As seen in Figure 3.4, sodium chloride is made up of the elements sodium and chlorine. Sodium has only one electron in its outermost electron shell. On the other hand, chlorine has seven electrons in its outer shell.

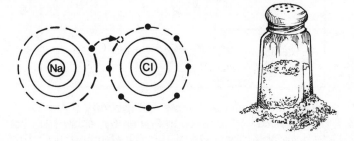

FIGURE 3.4 Sodium (Na) and Chlorine (Cl) Combine to Form Sodium Chloride (NaCl)

It is a rule of nature that elements strive to fill their outer shells with electrons. Therefore, it is only natural that sodium, with its one outer electron, will give it up to chlorine, an element that needs one electron to fill its outer shell.

13. The compound potassium chloride (KCl) is sometimes used as a substitute for table salt by people on a low-sodium diet. Referring to Figure 3.2, why might potassium combine with chlorine?

14. Magnesium oxide is made up of one atom of oxygen and one atom of magnesium. Therefore, its chemical formula is MgO. Referring to Figure 3.2, you will see that oxygen needs two electrons to fill its outer shell. Why might it combine with magnesium?

As shown in Figure 3.5, compounds can be made up of more than two atoms. Water, for example, is made up of two atoms of hydrogen and one atom of oxygen; its chemical formula is written H_2O.

15. In the formula for carbon dioxide (CO_2), what does the O_2 stand for?

Some compounds are quite complex. Many of the minerals in rocks, for example, are made up of fairly complex compounds. As

COMPOUND		FORMULA
Water	 H O H	H_2O
Mercuric oxide	 Hg O	HgO
Carbon dioxide	 O C O	CO_2

FIGURE 3.5 Some Simple Compounds (From *The Physical World, Second Edition* by Richard Brinckerhoff. Copyright © 1963 by Harcourt Brace Jovanovich, Inc. Reprinted by permission of the publisher.)

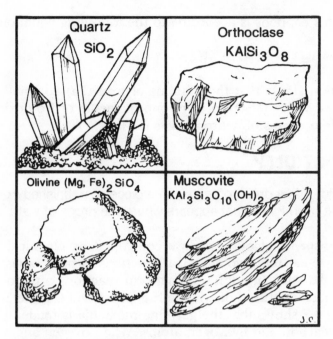

FIGURE 3.6   The Chemistry of Rocks

Figure 3.6 shows, many rocks have the element silicon in their chemical makeup. Such rocks are appropriately called *silicates*.

ACTIVITY

Go for a hike and collect five different rocks. When you get back, look up each rock type in a field guide to rocks and minerals. Note the chemical formulas of the minerals comprising each rock. You might try *A Field Guide to Rocks and Minerals,* by Frederick H. Pough (Peterson Field Guide Series, Houghton Mifflin Company, Boston 1976); or *Rocks and Minerals,* by Herbert S. Zim and Paul R. Shaffer (A Golden Guide; Golden Press [Western Pub. Co.], 1957).

CHEMICAL REACTIONS

Much of chemistry deals with how elements and compounds react with each other. Water and iron, for example, combine to form rust.

One way this happens is according to the following equation:

$$3Fe + 4H_2O \rightarrow Fe_3O_4 + 4H_2$$

This equation states that three atoms of iron (Fe) combine with four molecules of water (H_2O) to form one molecule of iron oxide (Fe_3O_4, rust) and four atoms of hydrogen gas (H_2).

ACTIVITY

Put a few drops of water on a piece of unfinished iron and let it sit overnight. What appears where the drops of water had been?

Many aspects of our lives are controlled by chemical reactions. The foods we eat undergo chemical reactions in our bodies, as do the medicines we take. In addition, chemical reactions often occur when an item is manufactured or when a fuel is burned (Figure 3.7).

Figure 3.7 shows that the gasoline molecule is made up of a long "chain" of carbon and hydrogen atoms. When oxygen and a spark are added to the gas, a very rapid reaction occurs in which atoms of oxygen join with hydrogen atoms to form water.

16. In the reaction shown in Figure 3.7, oxygen atoms also join with carbon atoms. What waste gas is produced with this reaction?

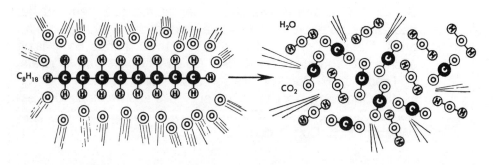

Gasoline + Oxygen + Heat ⟶ Carbon dioxide + Water + Energy

FIGURE 3.7   What Happens When Gasoline Is Burned (From *The Physical World, Second Edition* by Richard Brinckerhoff. Copyright © 1963 by Harcourt Brace Jovanovich, Inc. Reprinted by permission of the publisher.)

APPLIED CHEMISTRY

In the early days of chemistry, people believed that all things in the universe were made up of four basic elements—earth, air, fire, and water. Although this proved to be incorrect, people believed it throughout the Middle Ages. During the Middle Ages, it was not unusual for chemical experimenters, called *alchemists* (Figure 3.8), to attempt to turn ordinary metals into gold or to make "fountain-of-youth" potions.

Today chemists work in a field with many applications. They are employed in the mining industry, for example, discovering ways of extracting minerals from the surrounding ore. They also work in the oil industry, finding how fuels such as gasoline and kerosene can be separated from their parent crude oil.

Chemists also find work creating new products. Plastics, paints, and medicines, for example, were all created by chemists combining simple molecules into more complex molecules. The plastics used in skate boards, for example, were created by chemists, as were the shampoos we use to wash our hair. Let's examine how chemists apply their knowledge to some of the various aspects of our lives.

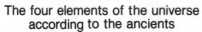

The four elements of the universe according to the ancients

An alchemist of the Middle Ages

FIGURE 3.8 Early Chemistry

Separation Chemistry

If someone gave you a beaker filled with sugar and sand, would you know how to separate the two? To do it grain by grain with tweezers would, of course, be very tedious and time-consuming. It would be much easier to pour water into the mixture and stir. The sugar would dissolve in the water but the sand would not. You could then pour off the water containing the dissolved sugar, leaving the sand behind. Then, if you boiled off the water in which the sugar was dissolved, you would be left with the original sugar.

17. Why wouldn't the procedure used to separate sand and sugar work with sugar and salt?

ACTIVITY

Separate a mixture of chalk dust and salt using the method described above.

In the mining industry, chemists work at separating one substance from another. That is, they separate the desired element from the ore it is found in. This is often done by first pulverizing the ore. In this way, separations can be done more easily. Iron, for example, is sometimes separated from its ore by passing a magnet over the pulverized ore. The iron, being magnetic, clings to the magnet, while the surrounding crushed rock dust does not.

Another useful way of separating one substance from another is to take advantage of their different boiling temperatures. A solution of water and alcohol, for example, can be separated in this way (Figure 3.9):

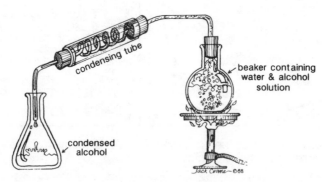

FIGURE 3.9 Separating Water and Alcohol

Alcohol boils at 78°C (172°F), whereas water boils at 100°C (212°F). When this alcohol/water solution is heated to 78°C, the alcohol begins to boil off. As Figure 3.9 shows, the alcohol can then be condensed and collected.

18. What is left in the original beaker after the alcohol has been boiled off?

Crude oil is a lot like a solution of alcohol and water. Instead of only two components, however, crude oil contains several liquids. Each of these liquids boils at a different temperature.

As you can see in Figure 3.10, the way the various liquids in crude oil are separated is the same method used to separate alcohol and water. As the temperature rises, each liquid becomes a gas at its given boiling temperature. This gas is then collected separately and condensed to produce that particular liquid, be it lubricating oil, fuel oil, kerosene, or gasoline.

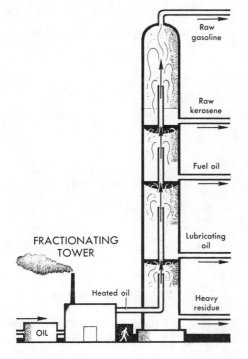

FIGURE 3.10 Separating Hydrocarbon Liquids from Crude Oil (From *The Physical World, Second Edition* by Richard Brinckerhoff. Copyright © 1963 by Harcourt Brace Jovanovich, Inc. Reprinted by permission of the publisher.)

ACTIVITY

You can separate oil and water without resorting to boiling. This is because oil simply floats on top of water. In a large pan of water, "create" an oil spill. Then figure out ways of containing it and cleaning it up.

Creating Compounds

To construct a house, building blocks such as lumber and bricks are used to make floors, walls, and ceilings. In chemistry, different elements are used as building blocks to make compounds. As you can see in Figure 3.11, chemists have learned to make quite a wide variety of compounds.

The compounds chemists work with can be as simple as one atom joining another, as is the case with NaCl—table salt (Figure 3.4). On the other hand, compounds may consist of thousands of atoms joined together to form a molecule of Teflon®, for example (Figure 3.12).

Notice in Figure 3.12 that the Teflon molecule is made up of a seemingly endless chain of carbon and fluorine atoms. Chains such as this may go on for thousands of atoms and are referred to as *polymers*. Polyethylene, a plastic used in products such as trash bags, is a polymer. Its chain structure is identical to that of Teflon. Unlike Teflon, however, polyethylene is made up of carbon atoms surrounded by *hydrogen* atoms.

FIGURE 3.11 Compounds Produced by Chemists

FIGURE 3.12   The Teflon Molecule

19. Diagram the structure of polyethelene.

Sometimes atoms will join each other quite readily to form compounds. This is the case with sodium chloride (NaCl) or water (H_2O), for example. In other cases, atoms do not combine so readily. In these cases, the chemist must rely on various "tools of the trade" to make sure various atoms link up as they are supposed to. One method of doing this is to use a catalyst. A *catalyst* is a material that helps launch a chemical reaction, but does not itself participate in that reaction. An example of a catalyst is in the catalytic converters of cars (Figure 3.13).

As you can see in Figure 3.13, exhaust gases from the engine are filtered through the catalytic converter and changed to carbon dioxide (CO_2) and water vapor (H_2O). The catalyst is commonly a mixture of two metallic elements, platinum and palladium. These elements, being the catalyst, are not changed during the purification reaction which occurs. However, small amounts of lead in the exhaust gases *will* cause the catalyst to react and become destroyed.

20. Why can't you use leaded gasoline in cars with catalytic converters?

Some of the work chemists do is with alloys. *Alloys* are mixtures of two or more metals. Bronze, for example, is an alloy of copper and tin. Alloys are not true compounds because the various metals do not combine chemically. They are a mixture, just as sugar and salt poured together constitute a mixture.

In their pure forms, many metals have certain undesirable features. For example, although gold is quite rare and pretty, it is a very

exhaust gases → → CO_2 + H_2O

FIGURE 3.13   A Catalytic Converter

soft metal. It is so soft in fact that you can actually stick your finger-nail into a pure sample of it! In such a form, it would be of little value in making jewelry, for instance. To increase its hardness, small amounts of copper or silver are added. In this way, a useful alloy of gold is created.

Alloys enjoy wide use today. In fact, most metallic objects are al-loys of some sort. Stainless steel, for example, is an alloy of iron, chromium, and carbon. Although composed principally of iron, stain-less steel will not rust like iron. Today, most knives, forks, and spoons are made of stainless steel. Before the invention of this alloy, most of these implements were made of more expensive silver.

ACTIVITY

Examine a sample of an element in its pure state. You can find such samples at gem and mineral shops, for example, or at mu-seums. How does the pure sample differ in appearance from something manufactured from the same element (such as a copper penny, or a gold ring)?

Detection Chemistry

Chemists who work in this field are detectives of a sort. They try to determine whether certain elements or compounds are present in a given substance. For example, in police work, they look for traces of poisons in foods or drinks, or they try to determine where a person has been based on microscopic substances found on that person's clothing. In other lines of work, chemists analyze drinking water for trace amounts of elements, such as those shown in Table 3.1.

Table 3.1 shows limits for various minerals in U.S. drinking-water supplies. A sample of water having one part per million of barium, for example, has one atom of barium present for every million molecules of water.

21. (a) Referring to Table 3.1, what is the recommended limit for lead in U.S. drinking-water supplies?

(b) What is the recommended limit for selenium?

22. Water flows through metal tubing in many water fountains. Wa-ter that sits in this tubing overnight will dissolve some of the metal, giving the water an unpleasant taste. If you have ever

TABLE 3.1   Limits for Minerals in Drinking Water

Mineral	Recommended Limit
Arsenic	0.05 parts per million
Barium	1.00 ppm
Cadmium	0.01 ppm
Chromium	0.05 ppm
Copper	1.00 ppm
Lead	0.05 ppm
Manganese	0.05 ppm
Mercury	0.02 ppm
Selenium	0.01 ppm
Silver	0.05 ppm

Compiled from EPA and U.S. Public Health Service standards.

been the first to drink from a water fountain in the morning, you have probably noticed this unpleasant taste. Why might running the water for a moment or two keep you from drinking water that contains high amounts of dissolved metals?

Environmental Chemistry

Many people believe that we are assaulting our planet faster than we are understanding it. We are able to create complex compounds such as pesticides and herbicides, for example, that kill off unwanted insects and weeds. Yet we are only now beginning to comprehend that some of these compounds are bad for our environment (Figure 3.14).

Environmental chemists examine some of the compounds our modern society produces. These scientists then try to understand how these compounds interact with their surroundings. If possible, they try to find a way of neutralizing them so that they will not pose a threat to the environment.

23. Chlorofluorocarbons are compounds that cause a destructive reaction in the ozone layer of our atmosphere. This ozone layer shields us from harmful ultraviolet light from the sun. What three elements are present in chlorofluorocarbons and give them their name?

FIGURE 3.14 Spraying Pesticide on a Vineyard (Photo Researchers, Inc.)

Acid rain is an environmental pollutant that affects lakes, rivers, and streams, causing fish and other living things to die. It is also thought to be responsible for dying forests in parts of Germany and elsewhere.

Acid rain is produced when water vapor in the air combines with gases given off when coal or oil are burned. Power plants, factories, and automobiles are major contributors to acid rain.

24. Canada receives acid rain that it claims originates in the United States. If the smokestacks are in the United States, why should Canada experience the problem?

SUMMARY

Chemistry played a major role in the House of Science because it provided many of the building materials used in its construction. Perhaps of greater importance, chemistry was also behind many of the products that are currently being used in the house. To understand the workings of this science, we must first know something about atoms, the building blocks of everything around us.

Atoms are made up of three principal particles—protons, neutrons, and electrons. The number of protons in the nucleus of the atom determines the type of atom it is. To date, 109 different kinds of atoms have been discovered. These account for the 109 elements listed on the Periodic Table of Elements (Figure 3.3).

The story of chemistry is told with the electrons that circle the nucleus of the atom. The number and arrangement of these electrons, particularly in the outermost shell, determines the chemical properties an element possesses. These chemical properties in turn determine how that element combines with other elements to form compounds.

When atoms join (and also when they split), we say that a chemical reaction has occurred. Many aspects of our lives are controlled by chemical reactions. These reactions occur when we eat food, for example, or when we burn a fuel such as gasoline.

Today, chemistry is applied in many ways. Chemists work at separating one substance from another, for example. They also work at creating the many products we see and use in our daily lives. In the area of detective work, chemistry is used to determine whether a poison is present in a given substance, or whether the fibers found on an article of clothing match those found at the scene of a crime. Finally, chemistry is used to examine the substances we use in our environment and to determine how these substances can be made safer.

25. The world's oceans pose a challenge to chemists of the future. Ocean water contains a vast abundance of useful minerals in solution, such as magnesium, silver, and gold. Finding an inexpensive way to separate these minerals would bring both fame and fortune to some aspiring chemist. Another challenge facing chemists is to find a way to produce fresh water from sea water inexpensively. Many believe that if this feat were accomplished, the world's food supply would double. Why do you think that is?

ANSWERS

1. 1

2. 92

3. Protons, neutrons, and electrons

4. (a) 2 electrons
   (b) 5 electrons
   (c) 8 electrons

5. shell 1—2 electrons
   shell 2—8 electrons
   shell 3—4 electrons

6. 5 electrons

7. 1

8. Helium and argon

9. Ca, Sr, Ba, and Ra

10. Lanthanide Series - Lutetium
Actinide Series - Lawrencium

11. 10

12. Goldwater

13. Like sodium, potassium has only one outer-shell electron (which it lends to the chlorine atom).

14. Magnesium has two outer-shell electrons to give up.

15. Two atoms of oxygen

16. Carbon dioxide

17. Both sugar and salt dissolve in water.

18. Water

19.

$$\ldots \overset{\displaystyle H}{\underset{\displaystyle H}{|}} \! - \! C \! - \! C \! - \! C \! - \! C \! - \! C \ldots$$

```
        H   H   H   H   H
        |   |   |   |   |
 ...  C - C - C - C - C  ...
        |   |   |   |   |
        H   H   H   H   H
```

20. The lead in the leaded gasoline would destroy the catalyst in the catalytic converter.

21. (a) 0.05 ppm
(b) 0.01 ppm

22. Running the water would "flush out" this undesirable water.

23. Chlorine (chloro), fluorine (fluoro), and carbon

24. The wind can carry these pollutants far from where they were produced.

25. Many present-day desert and semidesert areas could be irrigated to produce crops.

Chapter 4

PHYSICS

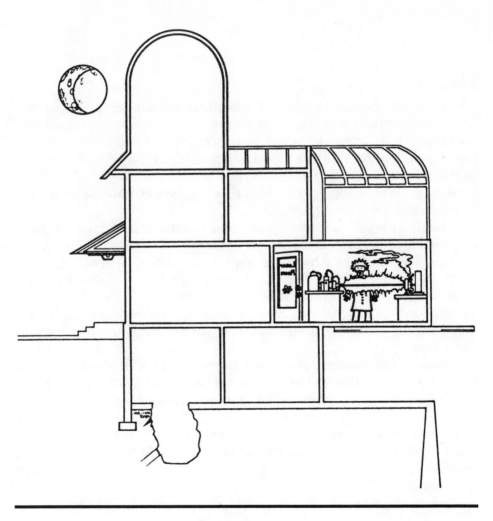

*The various principles of physics
helped to create the machines
used to construct the House of
Science. Motion, energy, and gravity
are but a few of the areas of study
of physics. Without physics, the
House of Science would be far
behind in construction, and things
we take for granted, such as
electricity, would not have been
discovered.*

The House of Science was built with the help of machines. Tools such as hammers, saws, and wheelbarrows are simple machines that helped make the work much easier. But more complex machines also aided in its construction. Power machines poured the cement for the foundation and cut the wood for framing the house. Power machines also transported the materials to the building site. Without these machines, construction of the house would have been a very difficult and time-consuming task.

Behind the various machines we use are the underlying principles of physics. How objects move, for example, is an area of study in physics, the result of which has been the creation of machines that carefully control motion. The saw, for example, is a machine that carefully controls motion in that it permits cutting in only the limited area desired. In similar fashion, more complex machines, although they may have more moving parts, carefully control the motion of each of those parts.

Energy, like motion, is another important area of physics. This is because to get motion in the first place, we must have energy. Energy is what powers the machines we operate. This energy can come in many forms, ranging from the mechanical energy our bodies provide to operate hand tools, to the electrical energy used to run the machines and appliances in the House of Science. To better understand physics, let's look at these two major fields it encompasses—motion and energy.

MOTION

We are all familiar with the concept of motion, for it can be seen all around us. Figure 4.1 shows some of the types of motion we observe in our daily lives.

1. Notice in Figure 4.1 that the water wheel harnesses the power of running water. What device (not shown in Figure 4.1) allows us to harness the power of the wind?

2. The flow of electrons through a wire is also a type of motion. This flow of electrons—electricity—is harnessed to power the various machines and appliances that we use. What two objects in Figure 4.1 are powered by electricity?

Factors Affecting Motion

Motion seems to be an easy enough concept to understand, but like many things that seem obvious at first, there's more to motion than meets the eye. Motion would be quite different on the moon or on a different planet, for example, because the pull of gravity would not be the same (Figure 4.2).

FIGURE 4.1   Some Types of Motion

61

FIGURE 4.2   Gravity on the Moon

Golf courses on the moon would have to be much longer because the moon's gravity is only one-sixth of the earth's. Thus, if a golfer could hit a ball 200 yards (183 m) on earth, this same golfer could hit the ball 1200 yards (1098 m) on the moon!

3. How far could this golfer hit the ball on a planet where the gravitational pull is twice that of the earth?

In the late 1600s, the English scientist Sir Isaac Newton showed that all objects, no matter how small, possess a gravitational pull. According to Newton, when we let go of a ball in midair, the earth's gravity acts to pull the ball to the ground. At the same time, however, the gravity of the ball, although much smaller, acts to pull the earth toward the ball!

Although the gravitational pull from any object on earth would be quite small, Newton reasoned that a large outside object, such as the moon, *would* exert a considerable pull on our planet. Evidence for this can be seen in the ocean tides. The cause of these tides is the moon's gravity actually pulling on the water in our oceans.

Newton found that the gravity from the earth is what keeps the moon in orbit around our planet. But if this were true, might gravity also be responsible for holding our earth (and the other planets) in orbit around the sun (Figure 4.3)?

Newton compared his calculations with actual observations made by astronomers and found that gravity indeed was the force holding our solar system together. Encouraged by this, he then speculated that the law of gravity operates not only in our own solar system, but throughout the universe as well, holding together clusters of stars and

FIGURE 4.3 Gravity Holds the Moon in Orbit around the Earth and the Earth in Orbit around the Sun

galaxies. To come full circle then, Newton reasoned that the same force that pulls a ball toward the earth also holds together the galaxies in our universe!

4. The force of gravity cannot be shielded by any known material. For example, a piece of thick slate (or any other material) will do nothing to shield the force of gravity coming from the earth. If you were to drop two objects, one over the piece of slate and the other beside it, would they fall at different speeds?

Although gravity plays a major role in the motion of objects, it does not play the only role. Other "players" enter the game also. For example, to get something to move, you must first overcome its *inertia*—that is, its tendency to stay at rest. The more an object weighs, the more inertia it has. A train car, for example, possesses far more inertia than a model replica of a train car. You could easily push the model but you would find it very difficult to move the real thing.

Inertia also keeps moving objects in motion. A moving ball, for example, tends to keep moving until it is slowed down by the air it is traveling through. In like fashion, a train car in motion tends to stay in motion.

5. The faster an object moves, the more inertia it possesses—that is, the harder it is to stop. Of course, the original weight (called *mass*) of the object is also important in determining its inertia. Which would be harder to stop, a ball moving at 50 miles per hour (80 km/hr) or a locomotive moving at 5 miles per hour (8 km/hr)?

The other major force affecting motion is *friction*. Friction tends to keep two objects from moving past one another. When you slide a book across the table, for example, it is friction that eventually brings it to rest. Figure 4.4 shows a closeup view of how friction works.

 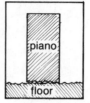

FIGURE 4.4 Friction

rough surfaces
catch on each other

As you can see in Figure 4.4, friction is caused by the roughness of two surfaces passing over each other. This roughness can be reduced by adding a lubricant such as oil to these surfaces. Such a lubricant acts to fill in the rough spots and thereby reduces the friction.

Various types of oil are used to lubricate the moving parts of machines. Without this oil, friction would cause these parts to heat up and eventually melt. This, of course, would ruin the machine. That is why you must keep oil in a car engine, for example, if you want it to last for any length of time.

6. When a road sign says "slippery when wet," what lubricant comes between the road surface and the wheels of your car?

Using Tools to Assist Motion

Now that you have learned something of how gravity, inertia, and friction affect motion, let's take a look at some of the tools used to assist us with this motion. As shown in Figure 4.5, even simple tools come in many shapes and sizes.

7. In Figure 4.5, the man is rolling the barrel up an inclined plane. What other picture is also an inclined plane?

8. A window fan is similar to what device shown in Figure 4.5?

ENERGY

The first section of this chapter covered motion—how things move and why. But to get motion in the first place, we need energy. Energy, like motion, is also a fundamental aspect of physics. This energy can come in many forms. Chemical energy is a form that comes from chemical reactions, such as those described in Chapter 3. *Kinetic energy*, the energy moving things possess, is perhaps the type of energy we are

FIGURE 4.5 Some Simple Tools

most familiar with. Stored energy is another form—a form called *potential energy*. Let's take a look at these and other forms of energy.

Potential and Kinetic Energy

Potential energy may be thought of as stored energy, while kinetic energy may be thought of as energy of movement. These two types of energy can be seen by looking at a roller coaster (Figure 4.6).

Looking at Figure 4.6, you see that at point A the roller coaster would be barely moving. But, despite its lack of movement, the coaster *does* possess quite a bit of stored energy at this point. This stored or

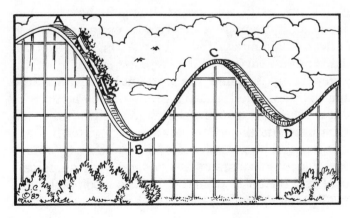

FIGURE 4.6
A Roller Coaster

potential energy it possesses is of course the ability to roll down the track to point B.

At point B, the potential energy has all been converted to motion or kinetic energy. Although the roller coaster is at the low spot, we know that it will not stay there because it possesses this energy of movement.

9. What type of energy does the roller coaster have at point C? At point D?

As in other machines, the motion of the roller coaster is carefully controlled so that it moves in only the desired direction. In this case, it is the track that keeps the roller coaster moving in the right direction.

10. What force prevents the roller coaster from going up and down hills indefinitely?

11. Potential energy can come in many forms. For example, it can be created by pumping water into a lake behind a dam. In this way, the energy is stored so that it can be used at later times, such as during times of low river flow. Then, it is used to keep the turbine blades in the dam turning. As the water is used to turn these blades, what kind of energy is it being converted to?

Heat Energy

Heat energy comes from sources such as the sun, a furnace, or a campfire. The use of this type of energy can be traced far back into history. Cave dwellers were perhaps the first to recognize that the heat from a fire was useful for cooking meat or for keeping warm.

Later in human history, as people began living in houses, they came up with the idea of an indoor fireplace. This invention contained the fire to a particular area, namely, the fireplace (Figure 4.7). Although this was perhaps not an earth-shaking step, the idea of containing a fire was very important. Because the next major advance, the idea of heating water in a confined area to produce steam, *was* an earth-shaking event. It produced the steam engine, and with it the machines of the Industrial Revolution!

12. In the heyday of locomotive steam engines, refueling stops carried both coal and water. In what part of the locomotive was the water put? (See Figure 4.7.)

66

FIGURE 4.7   Taming the Fire

After the Industrial Revolution was well under way, a new method of using fire was developed. This new invention was the internal combustion engine. With *internal combustion*, the fire is contained in the actual cylinders of the engine. As shown in Figure 4.8, an internal combustion engine operates by exploding very flammable vapors, such as from gasoline or alcohol. These exploding vapors in turn cause pistons to move up and down. This up-and-down motion is then converted into circular motion—motion that is used to turn the wheels of a train or car.

13. Notice in Figure 4.8 that the gas vapors are first compressed, then exploded with the aid of a spark plug. During which stroke does the explosion occur?

14. What device compresses the gases in the cylinder?

In looking at heat energy, it is important to remember that all things—be they solid, liquid, or gas—are made up of atoms and

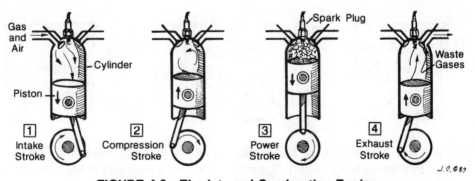

FIGURE 4.8   The Internal Combustion Engine

combinations of atoms called *molecules*. These molecules vibrate back and forth and collide with each other (Figure 4.9). In solids, the collisions are very ordered and do not result in any significant mixing of the molecules. In liquids, however, mixing does occur as molecules collide and slide past one another. Finally, in gases, the collisions between molecules are quite rapid, resulting in rapid mixing of the molecules.

15. When you put a spoonful of sugar in a glass of iced tea, you could simply wait for the dissolved sugar to mix by itself. Due to the motion of molecules, an even distribution of sugar would eventually occur in the glass. Because liquid molecules mix slowly, however, it might take days or even weeks for such mixing to occur. How do we greatly speed up the mixing process?

Temperature is a measure of a substance's heat energy. You could say that temperature measures how fast the molecules in that substance are moving. In other words, the faster the molecules move, the higher the temperature. To illustrate this, let's look at a substance familiar to all of us—water.

Water exists in three states: as a solid (ice), as a liquid (water), and as a gas (steam). In its solid form as ice, the molecules are ordered in a definite crystal arrangement. Although these molecules may vibrate, they are still under lock and key, so to speak. That is, they are not free to wander. With increasing heat, however, these molecules *do* break loose and are free to move about. When this happens, the ice turns to water.

If we were to continue heating our sample of water, the molecules would move faster and faster. Eventually, these molecules would move so fast that they could no longer be contained in the liquid and would

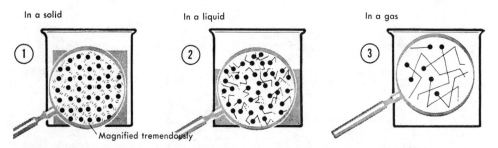

FIGURE 4.9 The Motion of Molecules (From *The Physical World, Second Edition* by Richard Brinckerhoff. Copyright © 1963 by Harcourt Brace Jovanovich, Inc. Reprinted by permission of the publisher.)

begin to fly around as a gas. We would notice this gas as the steam given off from the boiling water.

ACTIVITY

Place some ice cubes and a thermometer in a glass beaker. Slowly heat the beaker until the ice turns to water and eventually boils. Record the temperature every 30 seconds, then make a graph of temperature versus time.

Light Energy

Light energy comes from such sources as burning candles, electric lights, and the sun. This type of energy travels in waves, similar to the waves you can make when you jerk a piece of rope or throw a stone into a pond (Figure 4.10).

As shown in Figure 4.10, the girl can cause large waves or small waves, depending on how hard she jerks the rope or how large a stone she throws into the pond. It is important to note that the waves in the pond will all move outward at the same speed, no matter how big they are. Likewise, the waves created in the rope will all travel at the same speed.

ACTIVITY

Create your own waves by using a rope or by dropping pebbles into a tub of water. How can you change the size of the waves you create?

FIGURE 4.10 Creating Waves

69

Light energy also acts like a wave. However, it travels in waves that are much, much smaller than the waves in a rope or in water. In addition, these light waves travel at a very high speed—close to 186,000 miles per second (300,00km/sec.)! This speed is commonly referred to as the *speed of light*.

16. Given that the circumference of the earth is about 25,000 miles (40,000 km), about how many times would light travel around the world in one second?

The waves of energy shown in Figure 4.11 are part electrical and part magnetic—hence the name *electromagnetic*. Notice on Figure 4.11 that the waves get smaller to the left side of the diagram and larger to the right side. As a reminder, think of these waves as carrying energy of different wavelengths, just as different jolts to a rope carry energy of different wavelengths.

As you can see in Figure 4.11, visible light is only a small portion of the spectrum of electromagnetic radiation. Much of the energy we receive from the sun is in this small visible portion. However, a significant part of the sun's energy is in wavelengths invisible to us. It still exists, though. Heat, for example, is electromagnetic radiation that is largely in the infrared portion of the spectrum (Figure 4.11).

17. Ultraviolet waves can give us a sunburn, even though they are not visible to us. As seen in Figure 4.11, are these waves longer or shorter in wavelength than visible light?

18. We have harnessed many of the waves in the electromagnetic spectrum. For example, some waves travel through the soft tissues of our bodies but not through our bones. Doctors find these waves useful in setting broken bones, for instance. What are these waves called?

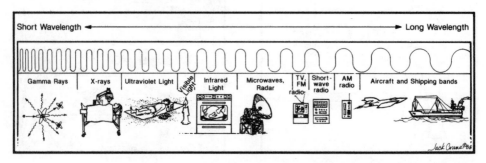

FIGURE 4.11   The Electromagnetic Spectrum

19. Microwave ovens make use of microwaves to cook food. Name another use of waves in this portion of the spectrum.

ACTIVITY

With the aid of a prism, separate visible light into its "rainbow" of colors. Each of these colors represents slightly different wavelengths of electromagnetic radiation. Using crayon, magic marker, or colored pencil, trace or recreate this visible spectrum on paper. Is the order of colors always the same?

20. As shown in Figure 4.11, light energy is only the *visible* portion of a larger source of energy called electromagnetic radiation. Name one type of electromagnetic radiation that exists but cannot be seen with the human eye?

Sound Energy

It is said that we live beneath an "ocean" of air. When this air moves, it sends out waves that our ears pick up and convert to sound. For example, striking a tuning fork causes the air directly adjacent to it to vibrate (Figure 4.12). These vibrations (waves) travel out in all directions, with some of them reaching our ears.

Sound needs a medium to travel through. We commonly think of sound as traveling through the medium of air. But sound travels about four times faster through water, a liquid medium, and through solids, the speed of sound is even faster!

21. Unlike the earth, the moon has no atmosphere. This means that if you were to take a tape player to the moon and turn it up to full volume, you would still hear nothing. Why not?

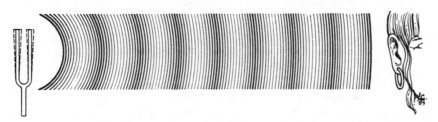

FIGURE 4.12 **Sound Waves from a Tuning Fork**

22. Sound travels at about 1,100 feet per second (335 m/sec) in air. This means that it takes about five seconds for it to travel one mile (3 seconds to travel 1 km). If, during a thunderstorm, you see a flash of lightning followed by thunder 15 seconds later, how far away was the lightning bolt?

23. If you see the lightning and hear the thunder from it at the same time, how far away was the lightning bolt?

ACTIVITY

During a thunderstorm, try to determine how far away lightning has struck using the rule given in question 22.

Electrical Energy

In 1820, Hans Christian Oerstead, while demonstrating electricity to a classroom of students, happened to place a magnetic compass underneath an electrified wire. To his surprise, the needle swung perpendicular to the direction of the wire (Figure 4.13)!

ACTIVITY

Connect a wire to one end of a dry cell battery. Then, place a compass under the wire and connect the wire's other end to the other terminal of the battery (as shown in Figure 4.13). What happens to the needle of the compass when you do this?

What Oerstead and his students had discovered was that a flow of electrons through a wire—electricity—will cause that wire to become a magnet. As such, the wire will deflect a magnetized compass needle.

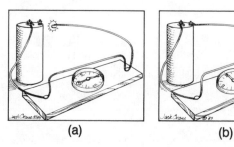

(a)   (b)

FIGURE 4.13   Oerstead's Compass Experiment

Ten years later, Joseph Henry and Michael Faraday sought to discover if the opposite of this might be true also. That is, could a moving magnet cause electrons to flow in a wire?

As shown in Figure 4.14(a), instead of moving cumbersome magnets, the researchers instead moved a wire in the presence of two magnets. They discovered that as long as the wire was kept in motion, an electric current would indeed flow in the wire!

It was only a matter of time before the findings of Faraday and Henry would be put to practical use. As shown in Figure 4.14(b), a water wheel can be made to keep a loop of wire in constant motion. As long as this loop of wire is kept moving between magnets, a current of electricity will flow through the wire. This flow of electricity can then be harnessed to do such things as light light bulbs, power appliances, or run computers. Thus, in looking back, the early experiments performed by Oerstead, Henry, and Faraday were the dawning of our age of electricity!

24. Looking at Figure 4.14(b), you can see that when a single loop of wire is rotated between two magnets, it causes an electric current to flow. This flow from a single wire is fairly weak, however. If many loops of wire are caused to rotate instead of a

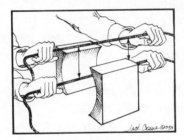

(a) by hand

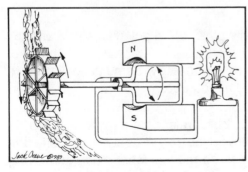

(b) by machine

FIGURE 4.14   Generating a Current

73

single loop, what effect do you think this has on the flow of electricity?

Modern generators use many loops of wire wound around a cylinder. This cylinder then rotates in the presence of powerful magnets, thus creating a strong electrical current. One way to constantly rotate this cylinder of wrapped wire is to funnel steam against the fan blades of a turbine, causing it to turn (instead of water, as shown in Figure 4.14b). This turbine is then connected to the cylinder of wrapped wire, causing it to turn and generate electricity.

25. To create steam to force against a turbine, many electricity-generating power plants need coal and water. What is the function of the coal? What is the function of the water?

Transformations of Energy

The different forms of energy discussed here can be transformed into one another. For example, when you burn a lump of coal, you have changed its chemical energy into heat energy. And, as we have just seen, this heat energy can be used to boil water to create steam—steam which in turn can be forced against the blades of a turbine, causing it to turn. Thus, the heat energy has been transformed into moving energy—that is, kinetic energy. This kinetic energy can then be used to turn loops of wire located between magnets, thus transforming itself into electrical energy.

26. Electrical energy is transformed into other kinds of energy in your home. For example, it can be used to power a radio, run a hair dryer, or provide illumination for a desk lamp. What type of energy is it being converted to in each of these three examples?

27. Some heat energy is usually given off in any transfer of energy. For example, when you turn on a light bulb, not all of the electrical energy is converted to light energy. Some is converted to heat energy as well. How can you determine this?

28. Chemical reactions in your body release energy when you digest food. This energy allows you to go about your daily life. When you use this energy to do manual labor, such as moving things, what type of energy is it being converted to?

NUCLEAR PHYSICS

In addition to motion and energy, modern physics also concerns itself with those building blocks of all things—atoms. We already know that atoms are made up of smaller particles, the most common of which are protons, neutrons, and electrons. But other subatomic particles exist as well, even though they are not yet well understood. Many of these particles have been given odd names, such as neutrinos, mesons, and quarks. It is thought that some of these particles hold the nucleus of the atom together.

Physicists currently study atomic particles by smashing atoms into each other. The particles given off from these collisions are then studied. The process is somewhat like a child taking a hammer to an unknown object in hopes of finding out more about it.

One atomic collision we are familiar with is when the unstable nucleus of the uranium atom is bombarded with neutrons. A "direct hit" causes the nucleus of this atom to split in two, releasing tremendous amounts of energy. This process is called *atomic fission* (Figure 4.15).

As seen in Figure 4.15, in the process of atomic fission, other neutrons are given off. These neutrons may then go on to split yet other uranium atoms. This *chain reaction*, as it is called, is the process used in nuclear energy plants. It is also the process that takes place when an atom bomb explodes.

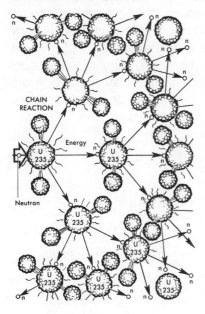

FIGURE 4.15 Atomic Fission (From *The Physical World, Second Edition* by Richard Brinckerhoff. Copyright © 1963 by Harcourt Brace Jovanovich, Inc. Reprinted by permission of the publisher.)

29. The discovery to atomic fission is an example of how science has changed the world we live in. Since World War II, for example, the citizens of our world have been living under a nuclear "trigger." What is meant by this?

SUMMARY

The House of Science was built with the help of machines. Behind these machines are the various principles of physics. Gravity, inertia, and friction, for example, are principles that deal with the motion of objects. By understanding principles such as these, we have been able to design machines that carefully control motion. That is, these machines permit motion in one direction only—the right direction!

To achieve motion, we need energy. Energy comes in many forms, such as heat energy, light energy, and electrical energy. These different forms are, in effect, interchangeable; that is, they can be transformed into one another.

Nuclear physics concerns itself with the structure of the atom. This branch of physics also looks at atomic reactions such as the fission of uranium.

30. Radioactive elements are elements in which the nucleus is unstable and, as a result, may break apart. When this happens, harmful radiation is given off. Uranium, thorium, and plutonium are all examples of radioactive elements. Radium also is a radioactive element—one that used to be put on watch dials. When atoms of this element broke apart, they struck a substance which in turn gave off light. In this way the watch dial could be seen in the dark. The use of radium on watch dials has since been banned. Why do you think this is?

ANSWERS

1. A windmill

2. The fan and the trolley car

3. 100 yards (91.5 m)

4. No

5. The locomotive

6. Water

7. The winding mountain road

8. The propeller

9. Point C—potential energy
   Point D—kinetic energy

10. Friction

11. Kinetic energy

12. The boiler

13. 3—the power stroke

14. The piston

15. By stirring the tea

16. 186,000 miles per second/25,000 miles = *about 7½ times* (300,000 km per second/40,000 km = 7½ *times*)

17. Shorter

18. X-rays

19. Radar

20. Many possible answers (see Figure 4.11)

21. There would be no medium to carry the sound to your ears.

22. About 3 miles (5 km)

23. It was very close!

24. Many loops of wire will increase the flow of electricity.

25. The coal is burned to heat the water and turn it to steam.

26. Radio—sound energy
    Hair dryer—heat energy
    Lamp—light energy

27. The light bulb gets warm.

28. Kinetic energy

29. We live in a world where nuclear weapons can be launched at a moment's notice—similar to pulling the trigger on a gun.

30. The radium gave off harmful radiation.

Chapter 5

GEOLOGY

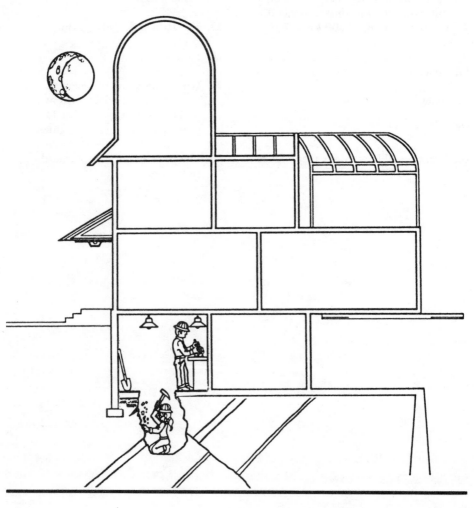

Located in the basement of the House of Science, geology is the study of the earth. Geology tells the history of our planet, what its interior is made of, and how its various landscapes formed.

T he science of geology deals with the study of the earth. As such, this science is appropriately located in the basement of the House of Science. From this vantage point, geologists are able to study the rock formations that underlie the house. They may look for coal and oil which, of course, could be used to heat the house. Or they may look for fossils that reveal something about the past history of the area where the house now stands. Geologists are not confined to working in the basement, however, for many of them work out-doors examining such things as how nature forms the landscapes we see. Others study volcanoes and earthquakes and how these phenom-ena might affect us in the future. To better understand the science of geology, let's begin by looking at the history of the earth.

HISTORY OF THE EARTH

The story of our earth begins about 4.6 billion years ago—a time when the rest of the planets in our solar system were also forming. Early in its history, our planet was made up of a hodgepodge of materials, some of which were radioactive. These radioactive materials gener-ated great quantities of heat, as did the large number of meteors that struck the earth at this time. The heat generated was sufficient to melt most of the planet, making our early earth a forbidding molten world!

About four billion years ago, the earth had cooled enough so that pieces of solid crust could form on the surface. Although a "skin" formed over the planet, molten material and gases continued to erupt from beneath the surface. These gases from the interior contained large amounts of water vapor that condensed in the atmosphere and fell as rain (Figure 5.1). However, the earth was still so hot that this rainfall instantly turned to steam upon striking the ground. With the passage of time, the crust of the earth cooled enough so that water *could* gather on the surface. Eventually, enough water accumulated to form the vast oceans of our planet.

Life is thought to have begun on earth about a billion years after our planet formed. This early life was of only the simplest sort, consisting for the most part of single-celled organisms such as bacteria. It was not until nearly 600 million years ago that more complex forms of life appeared. Some of these complex organisms had hard outer shells that were preserved as fossils. Indeed, it is from these fossil remains that we know of the existence of most of these organisms.

Geologists commonly divide the history of the earth into four different eras. As you can seen on Figure 5.2, the Precambrian Era was by far the longest of these eras. It also had the simplest life forms.

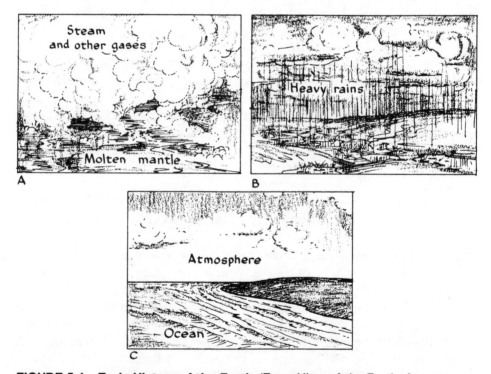

FIGURE 5.1 Early History of the Earth (From *View of the Earth, An Introduction to Geology* by John J. Fagan. Copyright © 1965 by Holt, Rinehart and Winston, Inc. Reprinted by permission of the publisher.)

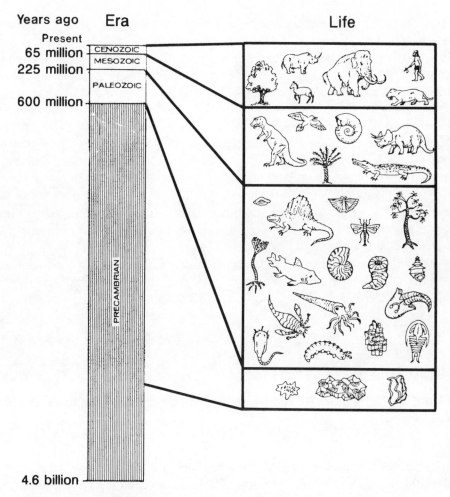

Years ago   Era   Life

FIGURE 5.2   The Four Eras of Earth History (Adapted from *Texas Fossils, An Amateur Collector's Handbook* by W.H. Matthews, III. Austin: University of Texas, Bureau of Economic Geology, 1960, Guidebook 2, plate 1.)

1. How many years did each of the four eras last?

2. When did the first land plants appear?

3. During what era did the dinosaurs inhabit the earth?

4. During what era did human life appear?

FIGURE 5.3 Life in an Early Paleozoic Sea (Smithsonian Institution Photo No. MNH653)

Figures 5.3 through 5.7 show what life may have looked like in the different eras of earth's history. These drawings were created by artists.

Many of the organisms shown in Figure 5.3 have since become extinct. That is, they have died out and have no living heirs. Others, such as the coral shown in the left foreground, *have* survived through the ages so that today we see their living descendants.

By mid-Paleozoic time, plants had begun to colonize the land. Trees that lived during this time resembled giant ferns (Figure 5.4). Insects and arachnids such as spiders also evolved during this time and were probably the first animals to live on land.

5. Figure 5.4 is a sketch of a mid-Paleozoic landscape. Notice that it does not show any large animals. Similar reconstructions of this time period may show insects or spiders, yet they also do not show large animals on land. Why not?

FIGURE 5.4   A Landscape in mid-Paleozoic Time

In late Paleozoic time, large land areas were covered by swamps (Figure 5.5). The trees that grew in these swamps looked quite different from those of today. A wide variety of animals had taken to the land by this time, and some insects, such as one that resembled today's dragonfly, had a wingspan of over two feet!

6. Generations of plants lived and died in the swamps of late Paleozoic time. In so doing, they built up a thick layer of material called *peat.* (Peat very closely resembles the garden mulch we know of as peat moss.) If conditions were right, a layer of peat may have been further compacted until it became as hard as rock. We mine this "rock" today and burn it for fuel. What is it?

Many organisms became extinct at the end of the Paleozoic Era, 225 million years ago. It is believed that the eruption of many volcanoes, along with intense mountain building, served to change climates dramatically on the planet. Many of the plants and animals simply could not adapt and as a result became extinct. Those that did adapt survived into the Mesozoic Era and evolved into life forms such as those shown in Figure 5.6.

FIGURE 5.5 **A Landscape in late Paleozoic Time** (Peabody Museum of Natural History, Yale University. Painted by Rudolph F. Zallinger)

FIGURE 5.6 **A Landscape in the Mesozoic Era** (Peabody Museum of Natural History, Yale University. Painted by Rudolph F. Zallinger)

7. The dinosaur Triceratops (left foreground, Figure 5.6) had what sort of protection for defending itself against attackers?

8. In Figure 5.6, what evidence do you see that the atmosphere in this area may be changing?

About 65 million years ago, mass extinctions again occurred on earth. Perhaps the best known casualties of these extinctions were the dinosaurs. All of the dinosaurs died out at this time. Recent evidence suggests that these extinctions were caused by an asteroid that collided with the earth, thereby changing climates. Whatever the cause, many plant and animal species were unable to adapt and died out. Those species that *did* survive lived to inherit the earth during the Cenozoic Era (Figure 5.7).

9. The Paleozoic Era ended when mountain building and volcano activity changed climates, resulting in the extinction of many species. The end of the Mesozoic Era also saw extinctions on a large scale, also probably due to rapid climate changes. The human species has the power to bring the Cenozoic Era to a close. How?

THE INTERIOR OF THE EARTH

In addition to learning about the earth's history, geologists are also interested in what the planet's interior looks like. To figure this out requires real detective work. This is because we are not able to observe the interior of the earth firsthand. That is to say, we are not able

FIGURE 5.7 A Landscape in the Cenozoic Era

to travel deep into the earth to see what it is made of. Instead, we must content ourselves with indirect evidence that we gather here on the surface. Let's look at some of this indirect evidence.

Volcanoes

When a volcano erupts, it gives off gases and molten rock from the interior of the planet. Scientists found that much of this molten rock cools to form a dark-colored rock called *basalt*. Is this basalt then what the interior of the earth is made of? At first guess, we might think so. But other evidence exists also. Let's look at it before we draw conclusions.

Diamond Pipes

In some places of the world, notably South Africa and Siberia, rocks found near the surface show signs that they formed much deeper in the planet. Such is the case with diamond pipes (Figure 5.8).

We know that the rocks in diamond pipes formed deep in the earth because diamonds are found embedded in these rocks. As you may know, diamonds are the hardest natural substance known. They can be created artificially in a laboratory, but only with the aid of intense heat and pressure. In nature, such heat and pressure would not be found near the surface of the earth. It would be found only by going much deeper into the planet.

10. Why do you think pressure increases as you go deeper into the earth?

Geologists are interested in diamond pipes, not only for the diamonds they contain, but also for the rock that surrounds these diamonds. Much of this rock has a greenish color and is known as *peridotite*.

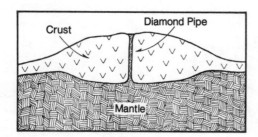

FIGURE 5.8 Cross-Section of a Diamond Pipe

You can discover an important characteristic of peridotite when you take a sample and heat it close to its melting point. As the rock partially melts, certain minerals in the rock melt before others. As a result, you first get a liquid that has a different composition than the original rock. In fact, this first liquid has a composition very close to that of basalt. And basalt, you may remember, is the rock found in many volcanic lava flows!

Could this similarity in composition between partially melted peridotite and the lava from volcanoes be just a coincidence? Many geologists don't believe so. They speculate that much of the interior of the earth is made up of the rock peridotite. They believe that, at a certain depth in the earth, temperatures and pressures are such that peridotite partially melts. The liquid formed from this melting then makes its way to the surface and erupts as the basaltic lava from a volcano.

11. (a) The concept of partially melting a rock can be demonstrated by putting a mixture of sand and water in the freezer. After a few hours, you will have a solid "rock" composed of sand and ice. If you then put this "rock" on the table at room temperature, it will partially melt. What will be the composition of the first liquid given off when this solid starts to melt?

(b) Would this liquid be of the same composition as the overall "rock"?

12. Chunks of solid peridotite sometimes appear in lava flows. Do you think this strengthens or weakens the theory that this lava originated from partial melting of peridotite? Why?

Earthquake Studies

By looking at volcanoes and diamond pipes, we have been able to establish evidence that the interior of the earth contains the rock peridotite. Studies of earthquakes give us another avenue for exploring the interior of our planet.

As shown in Figure 5.9, when a bell rings, shock waves are sent throughout the entire bell. Likewise, when an earthquake occurs, shock waves are sent throughout the entire earth. People usually feel these shock waves only a few miles from where the earthquake occurred. But sensitive instruments in the ground are able to detect them all around the world. These instruments are able to detect shock waves in Delaware, for example, from an earthquake that may have occurred in the Philippines, halfway around the world!

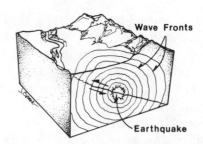

FIGURE 5.9   Shock Waves from a Bell and from an Earthquake

The shock waves given off from an earthquake are of two major types—*primary waves* (P-waves) and *secondary waves* (S-waves). These two waves are shown in Figure 5.10.

ACTIVITY

Demonstrate P- and S-waves for yourself by stretching out a Slinky. A sudden "push" on the Slinky® will create a P-wave, while shaking the Slinky back and forth will create an S-wave.

After an earthquake, P- and S-waves travel through the earth and are picked up by recorders at various places on the earth's surface. These recorders note not only that an earthquake has occurred, but also the types of waves it has received. For any given earthquake, the recorders might reveal the pattern shown in Figure 5.11.

Notice on Figure 5.11 that there is an area where no S-waves are picked up. It is known from laboratory experiments that S-waves do not

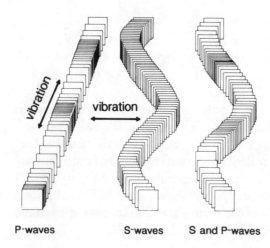

P-waves   S-waves   S and P-waves   **FIGURE 5.10   P- and S-Waves**

88

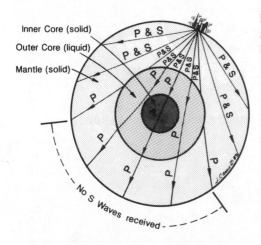

Inner Core (solid)

Outer Core (liquid)

Mantle (solid)

No S Waves received

FIGURE 5.11 The Passage of P- and S-Waves through the Earth Following an Earthquake

pass through liquids. Based on this, scientists believe there is a liquid core deep in the earth. This liquid core prevents the passage of S-waves.

Notice also on Figure 5.11 that a solid inner core has been drawn in the center of the earth. It is thought that this inner core exists because P-waves arrive "ahead of schedule" when traveling through that part of the earth. From laboratory experiments, we know that P-waves travel faster through solids than through liquids. Therefore, these faster arrival times for the P-waves can be explained by postulating a solid inner core to the earth.

13. Although S-waves travel through solids, they do not travel through the solid inner core of the earth. Why not?

Gravity Studies

Gravity studies allow us to further explore the interior of the earth. It is known, for example, that all objects exert a pull of gravity, however slight it may be. It is also known that the denser an object is, the greater its pull of gravity. A steel ball, for example, exerts a greater pull than a wood ball of similar size, although the gravity on either is insignificant because of their small size.

When dealing with an object the size of the earth, the pull of gravity *is* significant. And by looking at how the earth pulls nearby objects, such as the moon, asteroids, and satellites, we can get some idea of what the interior of our planet is made of.

As shown in Figure 5.12, a planet of low density, such as A, alters the path of a nearby object only slightly. Planet B, being made of

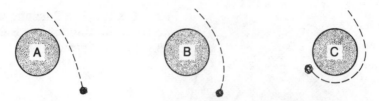

FIGURE 5.12   How Density Affects Gravitational Pull

heavier material, alters the path more than A. Finally, planet C, being made up of quite dense material, greatly alters the path of the passing object.

As with the planets shown in Figure 5.12, the earth's pull on nearby objects can be measured. From these measurements it has been determined that our planet, as a whole, has an average density of 5.5 grams per cubic centimeter. (By comparison, water has a density of 1 g/cm³). The earth's crust, on which we live, is known to have an average density of only 2.8 gm/cm³, while peridotite, the rock believed to comprise the mantle, has an average density of 4.4 gm/cm³. This means that if the earth as a whole has a density of 5.5 gm/cm³, the inner and outer cores, given their size, must be composed of material that is quite heavy. This material must have a density of 10 to 11 gm/cm³ to average out the relatively lighter crust and mantle. On earth, the only known material that has this required density is iron. Pure iron itself would make the earth a little too dense, but small amounts of impurities, such as nickel, would give it the required density.

14. Assuming the inner core and outer core of the earth are made of iron, would you weigh less or more than you do now if the earth's mantle were also made of iron instead of peridotite? Why?

By using these various lines of evidence—volcanoes, diamond pipes, earthquake studies, and gravity studies—we have been able to decipher what the interior of the earth probably looks like. The picture we come up with is shown in Figure 5.13.

15. The crust of the earth varies in thickness from about 3 miles (5 km) under some ocean areas to perhaps 100 miles (160 km) under the continents. Excluding this relatively thin layer of crust, what is the total diameter of the earth as shown in Figure 5.13?

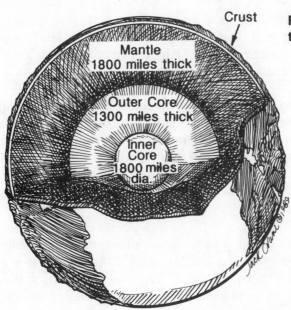

Crust

FIGURE 5.13 The Interior of the Earth

PLATE TECTONICS

You may have noticed that on a world map, South America and Africa look as though they could fit together like a jigsaw puzzle. In 1912, Alfred Wegener, a German meteorologist, proposed that these continents were indeed together at one time, then broke apart and drifted to their present-day locations (Figure 5.14).

Wegener based his theory of drifting continents not only on the similarity in fit between Africa and South America, but also on the resemblance of fossils found on each of these continents. He noticed, for example, that the remains of *Mesosaurus* (Figure 5.14) were discovered in both eastern South America and western Africa. This now extinct

 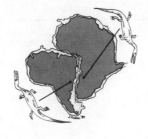

FIGURE 5.14 Alfred Wegener and His Theory

reptile is believed to have lived on land and in shallow waters close to land. Wegener pointed out that such an organism would clearly not have been able to cross an ocean separating these two continents. Thus, he postulated that the two continents were once together.

16. Many early geologists explained the similar fossil remains on South America and Africa by postulating a long narrow bridge of land that at one time spanned the Atlantic Ocean. Organisms such as *Mesosaurus* could have migrated between the two continents using this land bridge, according to these geologists. How did Alfred Wegener account for *Mesosaurus* being found on both continents?

To see how continents could move about, let's return for a moment to the mantle of the earth. The earth's *mantle* (Figure 5.13) is a layer of rock that is more rigid than steel. Yet, near the top of this mantle is a zone that is not as solid as the rest. In this zone, called the *zone of partial melting* (Figure 5.15), the mantle rock behaves in some ways like a very thick liquid. You could say that it has a consistency somewhat like tar.

ACTIVITY

Obtain some Silly Putty® and put a weight on it. Watch how this weight slowly deforms the putty. The putty is in some ways like the zone of partial melting within the mantle of the earth.

Overlying the zone of partial melting is the remainder of the solid mantle and the equally solid crust. What we have then is a rigid layer of rock lying over a semifluid layer in the earth's mantle. It is in some ways like the scum on top of a cup of cocoa overlying the liquid below.

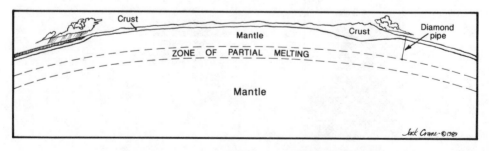

FIGURE 5.15   The Zone of Partial Melting

Just as the scum on top of cocoa migrates as you blow on it, the rigid upper layer of the earth migrates over the zone of partial melting. In this case, slow currents in the zone of partial melting are thought to cause the movement (Figure 5.16).

The upper layer of rigid rock on the earth is not one continuous piece. Rather, it is broken in several places so that various *plates* are formed. As you can see in Figure 5.16, one of these breaks occurs at the Mid-Atlantic Ridge. Here, new plate material is being formed as volcanoes bring up melted rock from below.

17. Another break between plates occurs just west of South America, where one plate is "diving" beneath the other. By looking at Figure 5.16, name the feature formed where this is happening.

In all, seven major plates form the top rigid layer of the earth (Figure 5.17). These plates all slowly drift over the zone of partial melting, much like "plates" of scum would drift over the surface of a cup of cocoa.

18. Mountains often occur where two plates collide. The Himalaya Mountains north of India, for example, are the highest mountains in the world. As seen in Figure 5.17, they are the result of the collision of what two plates?

When two plates collide, one is sometimes pushed beneath the other. As shown in Figure 5.18, the downturned plate eventually begins to melt due to the heat from the interior of the earth. This melted liquid, being lighter in weight than the surrounding rock, can sometimes find its way to the surface to erupt as lava from a volcano.

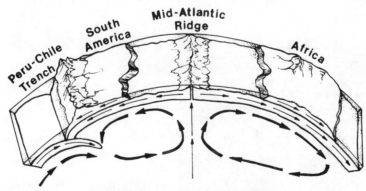

FIGURE 5.16 Migrating Plates

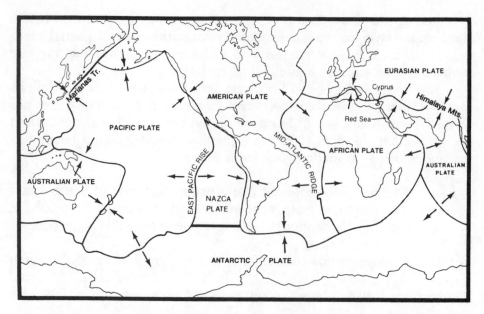

FIGURE 5.17   The Major Plates (From *Ecoscience: Population, Resources, Environment* by Paul R. Ehrlich et al. Copyright © 1970, 1972, 1977 by W.H. Freeman and Company. Reprinted with permission.)

A trench is often formed where one plate dives beneath another. In Figure 5.18, the trench is located on the ocean bottom, directly beneath the ship. The Marianas trench off the coast of Japan is an example of such a trench. Here, the ocean's depth—nearly seven miles (11 km)—is greater than anywhere else in the world! This trench is

FIGURE 5.18   The Collision of Two Plates

94

caused by the collision of the Pacific plate and the Euransian plate (see Figure 5.17).

19. As shown in Figure 5.16, the Peru-Chile trench parallels the west coast of South America. This trench formed as a result of part of the Nazca plate diving beneath the American plate. The Andes mountains of western South America also run parallel to this trench. Referring to Figure 5.18, how do you think they formed?

20. To make up for plate material lost at trenches, new plate material is formed at spreading zones such as the Mid-Atlantic Ridge. From Figure 5.17, what is the name of the spreading zone that is contributing new plate material to both the Pacific and Nazca plates?

Today, much evidence has been uncovered that serves as proof of the plate tectonics theory. So much evidence has accumulated, in fact, that today plate tectonics is considered the unifying theory of geology. Many features on the earth's surface, such as mountains and trenches, can be explained within the plate tectonics framework. In addition, many other phenomena, such as the prevalence of earthquakes in certain areas, can be explained according to the theory of moving plates.

21. The moving plates travel over the zone of partial melting at very slow rates. Some of these plates move only a fraction of an inch a year, while others move several inches a year. Europe and North America are on plates that are moving away from each other at a rate of about one inch per year (2.5 cm/yr). When you are 84, how much farther apart will these two continents have moved during your lifetime?

LANDSCAPE FORMATION

The earth's surface is unique in that it is where solid rock encounters the forces of nature. These forces work to break down the rock, and in so doing, fashion the landscapes we see. Just as a sculptor uses tools to carve a block of stone into a work of art, nature employs "tools" to create the landscapes we see. The tools of nature are a little different than the ones a sculptor uses, however.

In this section, we will look at some of the "tools" nature uses to fashion a landscape. Keep in mind that these tools are used in different ways depending on such things as the rock type and climate in an area. The wind, for example, is only an effective tool in dry areas where there is little vegetation to hold things in place. Similarly, chemical action is a more effective tool in limestone areas, where the rock is more susceptible to chemical attack.

Frost Wedging

This tool of nature works on the principle that water expands when frozen. This is demonstrated whenever you leave a beverage in the freezer for too long. It expands and breaks the container it is in.

In an outside setting, as shown in Figure 5.19, water first seeps into small cracks in the rock. Then, if the temperature drops below freezing, this water turns to ice and expands, thereby wedging the cracks a little further apart. Repeated periods of freezing and thawing will eventually break up the rock entirely.

22. Frost wedging is only effective in areas where the temperature fluctuates above and below freezing. It is not effective in tropical areas, for example, because the temperature seldom, if ever, drops below freezing. Frost wedging is not effective in polar areas either. Why not?

Plants and Animals

Figure 5.19 shows how plant roots can pry rock apart, just as frost wedging can. Plant roots also secrete acids that chemically attack the rock. In addition to plants, certain animals such as ants, earthworms, and rodents expose new fragments of rock to the surface. They do so with their burrowing activities. These animals also produce wastes that chemically attack the rock.

23. In a city, concrete sidewalks can be considered to be a rock. Where in the sidewalk do small plants start to grow and thus begin breaking up the rock?

Chemical Action

In addition to the acids generated by plants and animals, rock is attacked in another chemical way. When rainwater combines with carbon dioxide in the atmosphere, it forms a very weak acid. Although

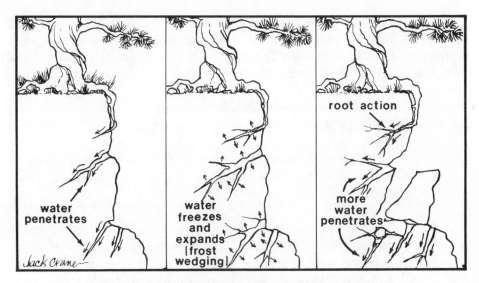

FIGURE 5.19   Frost Wedging and Root Action

this acid is weak, it will, over time, modify and dissolve rock that it comes into contact with. Over millions of years of time, for example, it is capable of transforming boulders of granite into clay!

ACTIVITY

Obtain some weak hydrochloric acid and put a drop or two on different types of rocks. Then test it on a sample of limestone. Which rock chemically alters the fastest? NOTE: **Hydrochloric acid can be dangerous. Do not use without adult supervision.**

Abrasion

This tool for breaking down rock is similar to using sandpaper or a file to wear something down. In nature, abrasion occurs when the wind blows sand grains against rocks, thereby "sandblasting" them. It also occurs when rock particles in a river abrade the bottom and sides of the channel. Probably the strongest abrasion, however, occurs underneath a slowly moving glacier, where ice and rock particles scour the underlying bedrock.

24. As a glacier overrides a landscape, it takes with it everything in its path, even large trees and other vegetation. What is the name given to the water coming off the leading edge of a glacier?

The "tools" described here for breaking down rock usually work together. Chemical action, for example, can only operate on the outer surface of a rock. But, with the help of frost action or the prying of tree roots, new rock surfaces are constantly being exposed to chemical action.

ACTIVITY

Go for a hike through a rocky area and note two or three instances of where tools of nature are at work breaking down the rock.

Once rock has broken down at the earth's surface, it can then be removed from the landscape. Just as a sculptor uses compressed air or a brush to remove loose material, nature employs its own tools for removing loose, weathered rock. Let's take a look at these tools.

Gravity

Avalanches and landslides are spectacular examples of how gravity can move rock material. Most weathered rock, however, moves by slower processes, such as *creep* and *slump* (Figure 5.20).

25. Notice on Figure 5.20 that creep gets its name because broken rock and soil near the top of the ground "creeps" downslope.

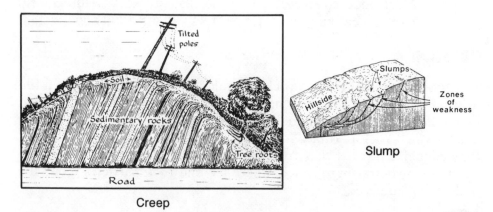

Creep

Slump

FIGURE 5.20 Creep and Slump Formation (From *View of the Earth, An Introduction to Geology* by John J. Fagan. Copyright © 1965 by Holt, Rinehart and Winston, Inc. Reprinted by permission of the publisher.)

The pull of gravity is responsible for this. How might the fence posts on a hill reflect the process of creep?

26. Slumps occur where a slope of material has become too steep to support itself. Why might heavy rains accelerate the formation of slumps?

Running Water

Running water is one of the most important tools nature uses to move weathered rock. On land having no vegetation, even the impact of a single raindrop is capable of moving material (Figure 5.21). In heavy rains, then, much material can be carried away as the water running off the ground joins to form rivulets, then streams, and finally, rivers.

27. Running water is a less effective tool in an area where there is extensive vegetation. Why do you think this is?

FIGURE 5.21 Raindrop Impact (U.S. Department of Agriculture, Soil Conservation Service)

Glaciers

Glaciers are a little bit like streams in that they usually occupy the valleys of a landscape. They are also like streams in that they flow. This flow—due to the gradual movement of ice—is usually on the order of only a few centimeters a day. As shown in Figure 5.22, as glaciers flow, they not only move broken rock, they also *create* broken rock by scouring the bedrock over which they move.

Areas that have been glaciated often leave a distinctive "fingerprint" on the landscape. For example, solitary boulders the size of a house are occasionally found in the middle of a field! The existence of these boulders testifies that the area was once overridden by a glacier capable of transporting boulders this size.

28. Very large boulders are sometimes found on the sea floor, far from any known source for these boulders. How might an iceberg—a piece of glacier broken off in the sea—be responsible for depositing these boulders?

Wind

The wind generally plays a minor role in removing weathered rock. It has its greatest effect in desert areas, where there are few plants to hold loose material in place.

In the 1930s, the wind was responsible for removing topsoil from farms in much of the midwestern United States. Poor farming practices, together with severe droughts, caused this area to be known as the "Dust Bowl."

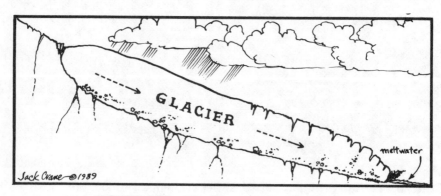

FIGURE 5.22   The Movement of a Glacier

29. As an area becomes drought-stricken, why does the wind become a more useful tool in the removal of material?

SUMMARY

Geology is practiced both inside and outside of the House of Science. The geologists who work in the basement of the house uncover rock formations in hopes of finding clues about our earth's past history. Geologists who work outside the house examine such things as how the forces of nature act to "sculpt" the landscapes we see.

Due to the work of geologists, we know that the earth's history can be divided into four major eras. Each of these eras saw new species of plants and animals inhabiting our planet. Also due to the work of geologists, we know that the interior of our earth is composed of an inner core, outer core, and mantle.

Looking closer at the earth's interior, we find that within the solid mantle there is a semifluid zone called the zone of partial melting. Above this zone is a layer of rigid rock that is broken into plates. According to the theory of plate tectonics, these plates migrate over the zone of partial melting, much like the scum on a cup of cocoa migrates across the surface. Today, many features on earth can be explained within the framework of plate tectonics.

In looking at landscapes, geologists examine the many "tools" nature uses as it sculpts the surface of our planet. The usefulness of these tools depends largely on the rock type and climate in an area.

30. Areas of land must first be created before the forces of nature can fashion them into landscapes. Keeping in mind the idea of moving plates (plate tectonics), name one way that new upland areas can be created.

ANSWERS

1. Precambrian: 4 billion years
   Paleozoic: 375 million years
   Mesozoic: 165 million years
   Cenozoic: 65 million years

2. Paleozoic

3. Mesozoic

4. Cenozoic

5. Large land animals had not yet appeared on the earth.

6. Coal

7. Horns (and a shield behind its head)

8. Volcanoes in the background are spewing gases into the atmosphere.

9. Nuclear war, pollution, overpopulation, etc.

10. The weight of overlying rock becomes greater.

11. (a) Water
    (b) No

12. It strengthens it. If partial melting occurs, you would expect to find pieces of the rock that is being melted.

13. They would first have to travel through the liquid outer core, which they cannot do.

14. You would weigh more. The overall earth would be much denser, making its pull of gravity stronger.

15. 8,000 miles (12,800 km)

16. He postulated that South America and Africa were once joined, so that species such as *Mesosaurus* could have roamed freely over both landmasses.

17. The Peru-Chile Trench

18. The Australian plate and the Eurasian plate

19. Part of the downturned Nazca plate melted, causing molten rock to come to the surface and erupt as volcanoes, thus forming the Andes mountains.

20. The East Pacific Rise

21. 84″/12″ per ft = 7 ft (2.1 m)

22. The temperature seldom gets *above* freezing.

23. In the cracks of the sidewalk

24. Meltwater

25. They may tilt downslope

26. They would lubricate the zones of weakness along which slumps can form.

27. The roots of plants tend to hold this material in place.

28. The boulders would have been in the iceberg when it was still part of the glacier. When the iceberg melted, it dropped these boulders on the sea floor.

29. Soil particles are no longer held together by moisture. Also, in severe drought the plants may die off, leaving no roots to hold the soil in place.

30. (a) By continents colliding with each other, or (b) by one plate diving beneath another, starting to melt, and thus bringing lava to volcanoes at the earth's surface.

Chapter 6

BIOLOGY

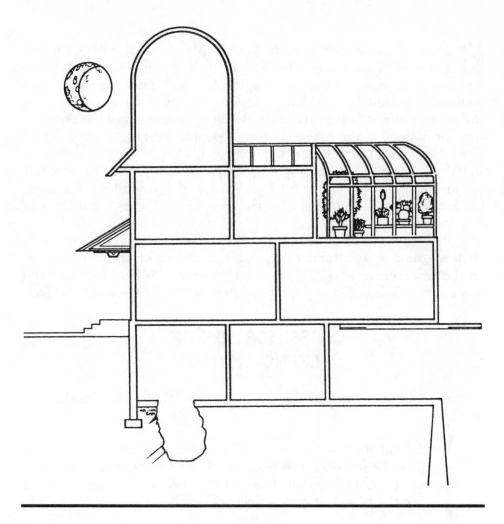

Biology has many settings in the House of Science. In addition to the greenhouse, biology also includes a study of the people, insects, animals, and grounds of the house. Biologists study ecology, evolution, and cells —the building blocks of all living things.

Biology is the science that deals with life. As such, visitors to the House of Science might be shown to the greenhouse as the biology room. In this setting, they could observe plants growing under controlled conditions of light, temperature, and moisture. But, as should be pointed out on the tour, the greenhouse is not the only setting for biology in the house. The very people working there are part of biology, because they are living beings also. Likewise, the occasional insect that finds its way into the house is a part of the science, because it too is struggling for life. Finally, the grounds surrounding the house are teeming with life, and therefore serve as a natural setting for the science of biology.

Biology overlaps with many of the other sciences. Geologists, chemists, and oceanographers, for example, all make use of biology, as do the researchers who study our environment. To better understand this science, let's begin by looking at how living things are classified.

CLASSIFICATION OF LIVING THINGS

Figure 6.1 shows some of the various forms of life that inhabit the earth.

1. Except for the mushroom (Fungi kingdom) and the seaweed (Protista kingdom, see Figure 6.2), the life forms shown in Figure 6.1 are either plants or animals. List, in two columns, which are plants and which are animals.

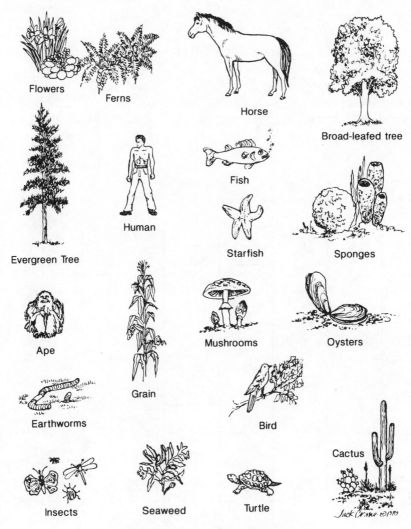

Flowers
Ferns
Horse
Broad-leafed tree
Evergreen Tree
Human
Fish
Starfish
Sponges
Ape
Grain
Mushrooms
Oysters
Earthworms
Bird
Cactus
Insects
Seaweed
Turtle

FIGURE 6.1 Living Things

To date, about a million and a half species of living organisms have been discovered on our planet! To distinguish these species from each other, we give them two names—much as you have two names that identify who you are. The two names we give to plants and animals are in Latin, however. That is because the man who devised the scheme, the Swedish biologist Carolus Linnaeus, decided on Latin as the language to work with. With his system, biologists the world over can communicate with each other, no matter what language they speak.

Under the classification scheme worked out by Linnaeus, organisms are grouped together based on their similarities. In this system,

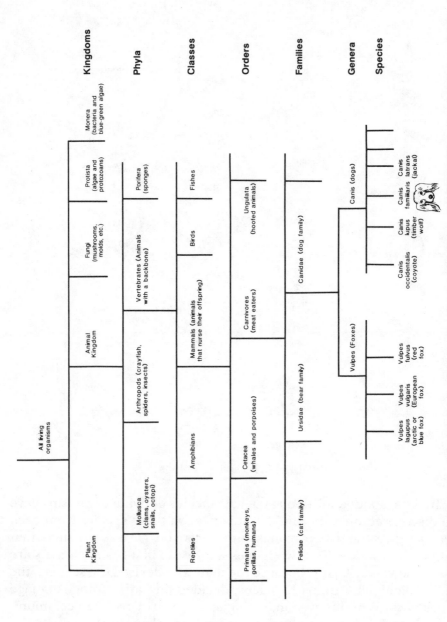

FIGURE 6.2 How the Family Dog (*Canis familiaris*) Fits into the Classification Scheme of Life

the family dog, for example, is given the name *Canis familiaris*, while its close cousin, the European wolf, is given the name *Canis lupus* (Figure 6.2). A more distant relative, the fox, is given the first name of *Vulpes* and a last name that depends on the type of fox it is. All of these animals, of course, are members of the dog family (Canidae). As such, they can be distinguished from members of the cat family (Felidae), such as lions, tigers, and domestic cats.

2. To which of three cats (lion, tiger, domestic cat) do you think the Latin name *Felis domestica* refers?

Although the dog and cat families are quite different, they are also similar in many ways. For example, they both are considered meat eaters and therefore can be classified with a larger order known as *carnivores* (Figure 6.2).

The classification scheme shown in Figure 6.2 is not complete because it shows detailed classification for only the dog family. Even so, it is not complete, for many other species of dogs and foxes could be named. In addition, many *varieties* of the domestic dog could be named, such as the cocker spaniel, Doberman, and so on. These varieties are not different species, however, because they are capable of breeding with each other, thereby producing mongrels (mutts).

As shown in Figure 6.2, human beings (*Homo sapiens*) are not part of the carnivore order that includes dogs and cats. But humans *do* nurse their young as cats and dogs do, and therefore are members of the larger class of animals known as *mammals*. And although mammals differ in many ways from birds, reptiles, amphibians, and fish, these animals all *do* have backbones and can therefore be grouped together as *vertebrates* (a phylum). Finally, vertebrates are members of a much larger group we know as the *animal kingdom*.

3. According to the classification scheme shown in Figure 6.2, what do the dog, bear, and cat families have in common?

4. What do the primates, cetacea, carnivores, and ungulata have in common?

5. (a) Reptiles, amphibians, mammals, birds, and fish are all part of a phylum of animals known as vertebrates. Why aren't insects a part of this vertebrate phylum?

(b) As shown in Figure 6.2, to which phylum do the insects belong?

6. Sponges are often mistaken for plants because they live attached to the sea bottom. To what phylum do these animals belong?

Some organisms, such as bacteria, fungi, and algae, cannot be classified as either plants or animals. That is because they have characteristics that make them different from members of either the plant or animal kingdom. Mushrooms, for example, grow like plants but do not utilize sunlight in building their structure as plants do. Mushrooms are therefore classified in the kingdom *Fungi* (Figure 6.2), along with such organisms as the molds that grow on food when it starts to decay.

Sometimes organisms cannot be classified as plants or animals because they exhibit features of both! These organisms are usually single-celled and are commonly classified in the kingdom *Protista* (Figure 6.2). The *Euglena,* for example, is a green microscopic organism that gathers sunlight and nutrients like a plant but that also can swim about like an animal.

ACTIVITY

Using the classification chart shown in Figure 6.2, play the game: "I am thinking of an animal." Have your friends ask you questions which require a yes or no answer only, such as "Is it a mammal?"

SURVIVAL MECHANISMS

Every plant and animal comes into the world equipped for survival. Animals such as turtles, lobsters, and clams are equipped with hard shells that offer them protection from their enemies. Others, such as deer and fish, are gifted with speed with which they can outdistance their enemies. Still other animals, such as monkeys, rely on their social organization to band together for survival.

7. What survival mechanism does the chameleon possess to protect it against natural enemies?

8. What protection does the skunk have against natural enemies?

9. Plants also come into the world equipped for survival. What sort of protection has the cactus plant against animals that might otherwise feed on it?

Sometimes protection is to be found in sheer numbers. For example, many insects are the prey to larger animals that feed on them. Although many of these insects get eaten, they survive from generation to generation because there are simply so many of them. Mosquitoes, flies, and gnats are all examples of insects that survive in this way.

10. Young tree seedlings serve as a food source for many types of animals. If so many of these seedlings are killed by grazing animals, how do you think these trees manage to survive from one generation to another?

As you have seen, survival often depends on being able to elude, repel, or outproduce enemies. But survival also depends on being able to make a living by gathering food. The various living organisms are well equipped for this task, too. Plants, for example, sink roots into the ground to gather water and nutrients from the soil, and animals such as birds possess beaks that enable them to gather the food they need.

Some plants and animals have clever and unique ways of gathering food. As seen in Figure 6.3, the ant lion is one of those species. In its desert environment, this animal digs a cone-shaped trap in the sand, then sits at the bottom waiting for an ant to fall in. When one does, the ant lion quickly goes to work undermining the sand, thereby causing a "landslide." Down with this landslide comes the captured ant. The ant lion is a poor sport, however, for if the ant should escape the landslide, it will use its flat head like a shovel to hurl sand grains at the fleeing ant!

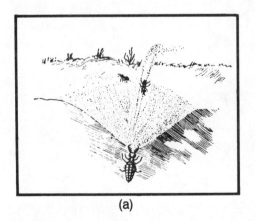
(a)

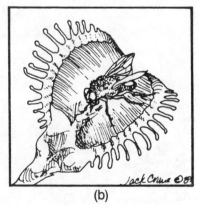
(b)

FIGURE 6.3 Two Unusual Methods of Gathering Food: (a) the ant lion and (b) Venus' flytrap.

The Venus' fly-trap plant shown in Figure 6.3 also has an unusual method of gathering food. This carnivorous plant has two hinged leaves that close on an unsuspecting insect such as a fly. The plant then digests its captured quarry before the leaves open again to catch another insect.

ECOLOGY

No one is an island, as the saying goes, and the same is true in the world of nature also. Living organisms all interact with each other in some way. Ecology is the study of how these organisms interact. It looks at what is known as the *web of life* (Figure 6.4).

Notice in Figure 6.4 that the web of life consists of both plants and animals. It also consists of microscopic organisms such as

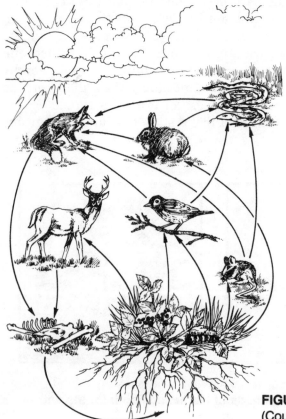

FIGURE 6.4 The Web of Life
(Courtesy Boy Scouts of America)

bacteria and fungi. These microscopic organisms act to decompose plant and animal remains. In this way, these remains are returned to the soil to be reused.

In the web of life, plants act as a principal means of support by providing food for animals. These plants take water and nutrients from the soil and air, then combine them with sunlight to produce new growth. This process is known as *photosynthesis.*

The plant foliage produced by photosynthesis serves as the primary food for many animals, including insects, rabbits, and deer (Figure 6.4). Other animals, such as the fox and snake, in turn feed on these animals that have fed on the plants.

11. The bird shown in Figure 6.4 feeds on plants both directly and indirectly. Explain.

The web of life differs in appearance from one area to another. Northern areas, for example, have caribou instead of deer feeding on plant life. The plant life is also different in appearance because of the climate. The basic relationships between plants and animals remain the same, however.

12. Algae are one of the major food producers in the sea. These microscopic organisms are the food source for many species of fish. In fact, it has been said that algae are to fish what grass is to a cow. What is meant by this?

13. The animals that feed directly on algae are equivalent to what four animals in Figure 6.4?

14. The shark is an animal that feeds on others but is not itself preyed upon. The shark is equivalent to what animal in Figure 6.4?

Habitat and Niche

Where an organism lives is referred to as its *habitat.* Examples of habitats include forests, grasslands, deserts, and oceans. The role an organism plays in its habitat is referred to as its *niche* (pronounced "nitch"). The niche of the rabbit in Figure 6.4, for example, is to feed on plants and in turn be fed upon by foxes and snakes. The niche of the deer is primarily to feed on plants.

15. What is the niche of the plants in Figure 6.4? That is, what function do they serve in the habitat?

16. What is the niche of the microscopic bacteria and fungi in the soil?

Interrelationships between organisms in a habitat are often quite complex and fascinating. For one thing, there are often a great many more niches than you might think. The rabbit shown in Figure 6.5, for example, can be subject to any of a number of parasites that live in or on its body. These parasites rely solely on the rabbit for their food supply.

17. The parasites shown in Figure 6.5 act to limit the size of the rabbit population. How?

18. In some habitats, organisms help each other in the battle for survival. The rhinoceros, for example, has parasites living on its back that serve as food for a certain type of bird. This bird benefits from the food supply and in turn warns the rhino of impending danger from predators. Other than signaling alarm, how else does this bird aid the rhino?

Many organisms change their role in the habitat from one season to another. A large number of plants, for example, only provide food in the summer months. This means that animals that depend on these plants for food must make provisions for the winter months. Insects do this by dying off in the cold weather and laying their eggs for the next

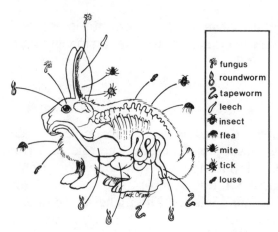

fungus
roundworm
tapeworm
leech
insect
flea
mite
tick
louse

FIGURE 6.5 Some of the Parasites that Attack a Rabbit

spring. Some birds adjust by migrating to warmer areas where the food supply is more plentiful. Animals such as deer simply make do on the decreased vegetation that is available.

19. How do bears adjust to the reduced food supply in winter?

Succession

Over a period of time, many habitats change with respect to the types of plants and animals that live there. This change is known as *succession*. Figure 6.6 shows how succession occurs on an abandoned field in a climate found in the southeastern United States.

Succession occurs because plants and animals change the environment in which they live. The first weeds and grasses in a bare field, for example, change the environment by shielding the hot, dry soil from direct sunlight. As these plants spread, the ground surface becomes cooler and more moist than it originally was. You could say that the environment at the ground surface has been changed. As a result of this change, certain plants such as shrubs can now begin to take hold and grow in the field.

As shrubs grow, they shade out the grasses and also add nutrients to the soil. In addition, they attract animals that ultimately enhance the soil. As a result of these changes, pine seedlings may soon take hold, and as they grow, they in turn shade out the shrubs. They are not able to shade out oak and hickory seedlings, however, which have now found the forest floor suitable. These seedlings will grow into large trees that eventually will shade out the pines.

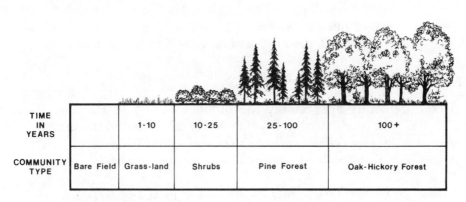

TIME IN YEARS		1-10	10-25	25-100	100+
COMMUNITY TYPE	Bare Field	Grass-land	Shrubs	Pine Forest	Oak-Hickory Forest

FIGURE 6.6 Plant Succession (Richards Rosen Press, New York)

20. Fire is common to some areas with a pine forest. In these areas, fire periodically sweeps through the underbrush but does not affect the pine trees because of their resistant bark. Why might fire prevent an oak-hickory forest from developing in an area?

The Human Factor

No discussion of ecology would be complete without covering the role of human beings in the environment. As a species, we can prey upon all other species but, in turn, are preyed upon by very few. Gone are the days when wild animals claim any significant toll on human life. The few organisms that do prey on us are the disease-causing parasites. Many of these have been controlled by modern medicine. Looking at our species, you could say that we occupy a unique place in the web of life. We not only are the top predator, we also can change or destroy the entire habitat!

21. The use of the gun has led to the decline of many species and the extinction of others. The passenger pigeon, for example, once numbered in the *billions* in North America. Today that bird is extinct due to overhunting. However, the gun is not the only way that animals can be destroyed. The bulldozer is equally effective. Yet most animals can flee from an oncoming bulldozer. Why, then, does this machine decrease their numbers?

22. As we come to better understand the web of life in an area, we should also develop a certain respect for it and treat it as if it were fragile, for in many cases it is. In Japan, for example, the English sparrow was deemed to be a nuisance and a hunting program was begun. The English sparrow's niche in the web of life was not considered, however, and as a result vast areas were soon plagued by insects. Why do you think this was?

In the past, we have been able to utilize the food web without destroying it. People have been fishing for centuries, for example, without depleting appreciably the number of fish in the sea (Figure 6.7). In recent years, however, we have greatly reduced the numbers of many fish species. What we have managed to do is come up with better and better methods of finding fewer and fewer fish!

23. In certain areas, why might it be a good idea to reduce or suspend fishing for a while?

FIGURE 6.7 Modern Fishing (Photo Researchers, Inc.)

EVOLUTION

Have the plants and animals we see today been the same throughout history? A careful study of nature reveals that they haven't. Death affects every living being, and so the "play of life" is carried on by the offspring of those living today. But are these offspring the same as their parents? In many respects they are. But just as you differ in appearance from your parents, the offspring of, say, a fish or a tree also differ slightly in appearance from their parents. And as sometimes happens, these small changes add up from one generation to another. With time, certain species may come to look quite different than their ancestors. This is the case with domestic dogs, for example (Figure 6.8).

All of the dogs shown in Figure 6.8 are descended from a single type of wild dog which probably resembled today's wolf. As early humans domesticated this animal, they also began a selective breeding program with it. That is, instead of allowing dogs to breed with a mate of their own choosing, as happens in nature, they instead chose the individuals that would breed. If large dogs were desired, for example, a large male and female were chosen to breed. It was thought (and correctly so) that the offspring of such a match-up would most likely grow to be larger than average. Generations of such breeding succeeded in producing large dogs such as today's Great Dane and German shepherd.

When we look at the wide variety of dogs there are today, it may seem hard to believe that all these breeds can trace their descent to a common type of ancestral wild dog—yet this almost certainly is the case. By artificially selecting which dogs will breed in each generation, we have come up with the wide variety of dogs shown in Figure 6.8.

FIGURE 6.8 Domestic Dogs (From Raymond C. Moore, *Introduction to Historical Geology*, 1958, McGraw-Hill Book Co.)

24. **(a)** Suppose breeders wanted to create small dogs with long snouts. Such dogs would be useful for flushing small animals out of their underground burrows. What characteristics would they have looked for in dogs that were to be bred?

(b) Name a modern-day dog that is the result of such breeding.

In 1859, Charles Darwin and Alfred Russel Wallace, two English naturalists, independently proposed the theory that all of earth's living organisms descended from a common ancestor. Just as today's breeds of dogs can be shown to have evolved from a common species of wild dog, these two researchers took the larger step of saying that *all* life on earth evolved from a common ancestral species. This species would probably have been a simple one-celled organism similar to present-day bacteria.

25. The evolution of the first two-celled organism may have been the result of a mutation, or fluke of nature. Once this organism appeared, however, it may have found that one cell was slightly better at gathering food, for example, while the other cell was more adept at moving the organism from place to place. What advantage would an organism have in being able to move from place to place in its environment?

26. As multi-celled organisms evolved, certain cells in their bodies came to perform certain specific tasks. In the higher animals, for example, certain cells control the functioning of the rest of the body parts. These cells are a part of what major organ?

Darwin formulated his theory of evolution based on a voyage he had taken around the world. During this voyage, he had the opportunity to observe plants and animals quite unlike those seen in his native England. He kept careful records of his observations and brought these records back with him when he returned from his trip. Let's look at some of the observations Darwin made, for they were to fit into the theory of evolution he was formulating:

Differences between individuals. Darwin recognized that individuals of a species differ from each other. In our own human species, for example, we can see differences in such things as body size and build, color of eyes, hair color, and so on. In plants, differences can be found in such things as height, overall shape, and the number of seeds produced. For any given plant or animal species, Darwin recognized that there would be differences between the individuals in that species.

Tendency toward overpopulation. Many species on earth produce a great number of offspring. If a high number of these offspring lived, that species would soon overrun the earth. Many trees, for example, produce thousands of seeds each year. If each of these seeds were to grow into a mature tree, it would not be long before that species of tree predominated on the planet. Another example is fish, which commonly lay millions of eggs at a time. If each of these eggs were to grow into a mature fish, that species would soon dominate the seas. Darwin recognized that even organisms that produce only one offspring per year, such as horses, are theoretically capable of overpopulating the earth, although not as rapidly.

Struggle for survival. Darwin recognized that there are limitations to the amount of land and resources available on our planet. There is only so much edible food, for example, just as there is only so much living space on land or in the sea. Because of these limitations, Darwin reasoned, all the plants and animals must compete for what *is* available. He stated that this creates a struggle for survival in which those that are most fit survive—hence the phrase "survival of the fittest."

27. Sometimes the organisms that win the struggle for survival are those that can get along the best. Many animals live in groups, for example. This increases an individual's chances for survival in many ways. Name a way you think this would increase its chances for survival.

Because only a small number of offspring can live to adulthood, there must be some way of choosing which ones will make it and which ones will not. Darwin reasoned that nature makes this choice, and thus referred to the process as *natural selection.* To illustrate natural selection, let's look at amphibians—animals that spend part of their lives in water and part on land (Figure 6.9).

Referring to Figure 6.9 (left side), there were times in the past history of the earth when shallow inland seas covered much of what today is land. As these seas began to dry up, they formed lakes. And as these lakes in turn began to go dry, the fish that lived in them faced the danger of extinction. But among those fish were some species that

FIGURE 6.9 Evolution of the First Amphibians (Neg. nos. 322871 and 322872, Courtesy Department of Library Services, American Museum of Natural History)

had developed primitive lungs for breathing out of water. And among these lungfish were individuals that had more muscular fins than the others. These individuals had the ability to crawl out of the water onto land where a vast food supply of plants and insects awaited them. Because these individuals were assured of a food supply in hard times, they were more likely to survive than the fish that were not able to leave the water. And as you know, those that survive are those that live to reproduce.

28. (a) We have discussed how breeding large dogs generally produces offspring that also grow to be large. What kind of fins (muscular or less muscular) do you think the offspring of surviving lungfish tended to have?

(b) As seen in Figure 6.9 (right side), did these lungfish change their appearance with time and thereby evolve?

Mutations sometimes decide the course that evolution will take. For example, frogs are occasionally born with four back legs instead of two. As with most mutations, however, these individuals have a very low chance of survival. Occasionally, though, a mutation occurs that increases an organism's chances for survival. This may have happened with an amphibian that was born with the ability to produce eggs that could be laid on land. Such an individual would not have had to return

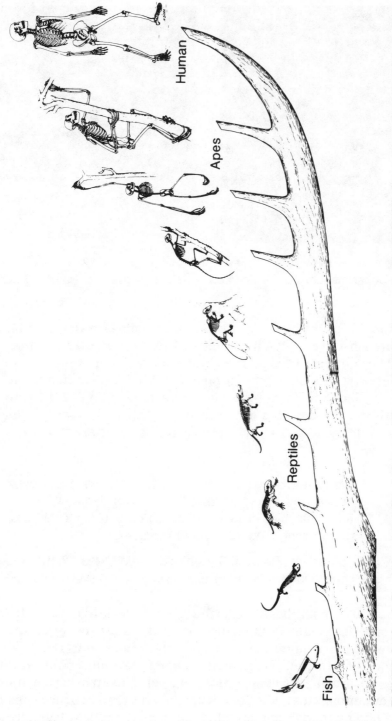

FIGURE 6.10 A Branch on the Evolutionary "Tree of Life" (Neg. nos. 311665, Courtesy Department of Library Services, American Museum of Natural History)

to water to lay its eggs. If this mutation was passed on to succeeding generations, the result would have been the creation of an animal no longer dependent on water for part of its life cycle. Such an animal could travel further inland and occupy new habitats. Such an animal would have resembled today's reptiles!

ACTIVITY

Plan three nature walks, each within a different habitat, such as a wooded area, an open field, and a stream or pond. Each time bring a note pad and pencil along and make a list of organisms you encounter along the way. Then, compare your three lists and identify those organisms that appear only on two but not all three lists. These organisms have adapted to two different habitats. Do any organisms appear on all three lists?

If evolution has indeed occurred, we should expect related organisms to look alike. That this is the case can be seen in Figure 6.10. This figure shows one branch on the evolutionary "tree of life"—the branch that includes our own human species.

Notice in Figure 6.10 that the apes and the human have many similar features. Arm and leg structures, for example, are remarkably similar, as are rib cage and backbone structures.

Scientists believe that the early ancestors of humans were apes. These apes lived, for the most part, in forests. With time, some of them began to move out onto grassland areas. Among the apes that did move onto the grasslands were a few individuals that had a more upright posture than the others. These apes found that they could move around better, and this greatly increased their chances for survival. Because these apes tended to survive, they were able to breed and pass on their upright posture to future generations. With time, the apes that lived in the grasslands became fully upright and acquired other features that better suited them to life in their new habitat. With time, they no longer looked like apes; instead they began to look like early humans.

29. In Figure 6.10, what major change can be seen in the skull from the early apes to the human?

If evolution has indeed occurred in the past, we should expect to find evidence for it in the fossil record. Fossils, you may recall, are the bones and other evidence of life (such as tracks) we unearth from time to time. The fossil remains for many species show dramatic changes

121

 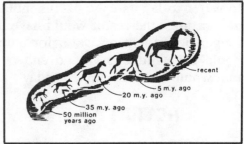

FIGURE 6.11 **The Evolution of Horses** (Right: Courtesy Boy Scouts of America)

over the generations. When looking at horses, for example, we can see that members of this family are much larger today than they were 50 million years ago (Figure 6.11).

30. Referring to Figure 6.11, why did natural selection favor the evolution of large horses over small ones?

When we look at evolution, we must view it against the backdrop of drifting continents. As geologists tell us, the crustal plates—of which the continents are a part—have probably drifted throughout much of the earth's history (Figure 6.12).

31. Fossils have been unearthed of organisms that lived in North America 225 million years ago. These fossils are of organisms that lived in a relatively warm climate. Later remains, however, indicate that many of these organisms evolved so that they

225 million years ago   65 million years ago   Present day

FIGURE 6.12 **The Drifting Continents**

could withstand colder and colder temperatures. Looking at Figure 6.12, why do you think the northward movement of North America caused the evolution of organisms better adapted to a colder climate?

32. Many organisms that freely roamed from North America to Africa 225 million years ago are not able to do so today. What happened between that time and today to prevent this?

Evolution is one of the underlying principles of biology, and a knowledge of how it works is important to many people. People who are in the business of breeding animals, for example, will breed only individuals with certain desired traits. Likewise, plant breeders will breed only plants with the highest yield, or plants whose flowers are the most unusual or pretty. In so doing, they are controlling the evolution of these species.

Evolution is also important to people who work at controlling insect populations. Here, we see it taking place quite rapidly—much to the dismay of those who make pesticides. When a pesticide is sprayed on a population of mosquitoes, for example, most of them are killed. There are often a few insects, however, that are not susceptible to the pesticide. These insects frequently pass on their resistance to their offspring so that, within a few generations, a new breed of mosquito has evolved. This new breed is, of course, resistant to the pesticide.

33. A knowledge of how evolution works has resulted in many of the crops we currently grow. In fact, some of these crops, such as corn, have evolved to such an extent that they can no longer survive in the wild by themselves. Their ability to scatter their own seeds, for example, has been lost in the process of breeding for larger ears of corn. Large ears of corn mean a higher yield to the farmer. Whether the corn plant can scatter its seeds from year to year in the wild is not a concern to farmers. Why not?

CELLULAR BIOLOGY

As you know, biology is the study of life. This life is made up of millions of small compartments called *cells*. Most of these cells are microscopic in size, but a few are much larger. Birds eggs, for example, are cells that are quite large in size.

123

Cells make up the living things we see around us. They also make up the tissue in our bodies. As shown in Figure 6.13, cells can vary considerably in their appearance.

Although cells vary widely in size and shape, they *do* share three basic components. As shown in Figure 6.13, the first of these components is the *nucleus,* a rounded ball in the interior of the cell. Surrounding the nucleus is a mass of material important to the operation of the cell. This material is referred to as the *cytoplasm.* Finally, at the outer edge of the cell is the *cell membrane*. The cell membrane is what holds the cell together.

ACTIVITY

Make sketches of different plant and animal cells under a microscope. Notice that many of these cells vary greatly in appearance. Under high power, look for the nucleus, cytoplasm, and cell membrane in particular cells.

In more complex organisms, cells group together to form tissues. Tissues in turn group together to form organs such as the heart or lungs. Organs, of course, become parts of systems, such as the circulatory system, nervous system, or skeletal system. All of these systems contain billions of cells as their building blocks. And as you can see in Figure 6.13, these specialized cells look quite different from each other, even though they all may be found in the same organism.

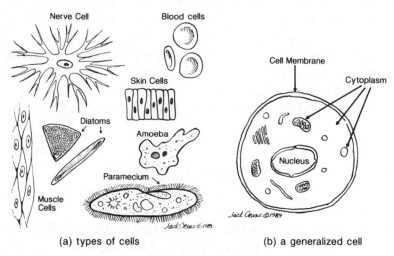

(a) types of cells   (b) a generalized cell

FIGURE 6.13   Cells

34. Although the cells in Figure 6.13 look quite different, they all contain what three basic components?

As seen in Figure 6.14, cells reproduce by dividing in half. This process of cell division (known as *mitosis*) is the way plants and animals grow larger and mature. It is also the process that heals you when you are injured.

In Figure 6.14, the three pairs of black lines in the cell are chromosomes. *Chromosomes* are responsible for the characteristics we inherit. Although three pairs of chromosomes are shown in this figure, this number varies from species to species. Human beings, for example, have 23 pairs of chromosomes.

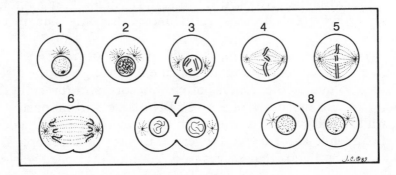

STEP 1 - The cell prior to reproduction.

STEP 2 - Black lines called chromosomes start to appear in the nucleus of the cell. These chromosomes contain the cell's genetic information.

STEP 3 - The chromosomes thicken until distinct pairs can be seen.

STEP 4 - The nucleus disappears and the chromosomes begin to align themselves in the center of the cell.

STEP 5 - Complete chromosome alignment in the center of the cell.

STEP 6 - The paired chromosomes divide, with one of each pair going in opposite directions.

STEP 7 - Two new nuclei appear and the chromosomes elongate and thus become less visible. The cell membrane also begins to divide.

STEP 8 - Reproduction of the cell is complete. Two cells have formed from one.

FIGURE 6.14   Reproduction of a Cell

125

35. Referring to Figure 6.14, during what step do the chromosomes actually divide?

Chromosomes are long molecules that have the shape of a twisted ladder (Figure 6.15). The major "ingredient" in these chromosomes is a substance referred to as *DNA* (short for deoxyribonucleic acid).

As shown in Figure 6.15, each rung of the DNA ladder is occupied by two molecules. Prior to cell division, these molecules separate from each other and the ladder unwinds. Each half then seeks molecules within the cell that will complete a new ladder. When this process is completed, the actual reproduction of the cell (Figure 6.14) can begin.

Notice in Figure 6.15 that only four types of molecules make up the rungs of the DNA structure. The names of these molecules are abbreviated with the letters A, C, G, and T. Notice that A molecules only pair up with T molecules, and C molecules only pair up with G molecules.

Incredible as it may seem, the wide variety of life we see in this world is due to the length and arrangement of rungs on the DNA ladder. It is also due to the *number* of chromosomes in the cell. Thus, a cell with 14 chromosomes of a given DNA code will be part of a pea plant, while a cell with 42 chromosomes and a different DNA code will be part of a white rat. Taken all together, it is this microscopic DNA in the nucleus of all cells that serves to direct the "orchestra" of life!

36. In Figure 6.15, when the new strands have completed their "shopping" for molecules, how will they resemble the original molecule?

Many one-celled organisms reproduce by the method of cell division shown in Figure 6.14. And, as mentioned before, this is also the

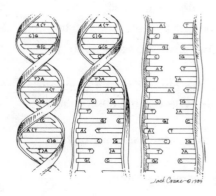

FIGURE 6.15   The DNA Molecule

method by which the higher organisms grow and repair themselves. Higher organisms do not reproduce by this method, however. Instead, they reproduce by means of specialized cells that their bodies produce. These specialized cells contain only half the chromosomes that are present in an ordinary cell. Such cells are referred to as *sperm cells* and *egg cells*. (The process by which they are formed is known as *meiosis*.)

37. **(a)** Human beings have 46 chromosomes in each of the cells of their bodies. There is one exception to this though. Male sperm cells and female egg cells each contain how many chromosomes?

(b) In humans, when a sperm cell joins an egg cell to begin a new life, how many chromosomes are present in the new cell?

38. In trees, what is the fertilized egg called, from which a new tree can grow?

How does a fertilized egg grow into a human baby, and not a chicken, a frog, or a pin oak? How does this developing fetus "know" what it is going to develop into? This is one of the mysteries that biologists are currently trying to solve. They believe that certain chemicals may interact with the DNA code in every cell to direct the development of the growing fetus.

39. Some fertilized eggs develop directly into typical individuals, whereas others produce embryos or larval stages. What is the swimming stage of a frog, before it is called a frog?

SUMMARY

A visitor to the House of Science might be shown to the greenhouse as the biology room. But life exists in other rooms of the house also, as well as on the grounds surrounding the house. This life is quite diverse, with 1.5 million different species having been discovered on the planet so far. These different species are classified according to characteristics they have in common with one another.

Every organism comes into the world equipped for survival against its enemies. It also enters the world equipped to gather food for itself. How these organisms in turn interact with each other in their environment is called *ecology*.

The theory of evolution accounts for how species change with time. According to evolution, the forces of nature decide which individuals will survive and pass on their traits to offspring.

Cellular biology deals with the building blocks of life—cells. It looks at such things as cell division, chromosomes, and the way cells reproduce themselves. This branch of biology also examines the growth and development of living organisms.

40. Genetic engineering is one field of biology that holds much promise in the future. In this field, biologists alter the DNA code in cells to try to cure dreaded diseases, for example, or to produce varieties of crops that are resistant to parasites. One recent accomplishment in this field was the development of certain crops that are resistant to frost. Why would these frost-resistant crops be an advantage to the farmer?

ANSWERS

1. | Plants | Animals |
   |---|---|
   | Flowers | Horse |
   | Ferns | Human |
   | Broad-leafed tree | Fish |
   | Evergreen tree | Starfish |
   | Grain | Sponges (see |
   | Cactus | Figure 6.2) |
   | | Ape |
   | | Earthworms |
   | | Insects |
   | | Oysters |
   | | Bird |
   | | Turtle |

2. The domestic cat

3. They are all carnivores.

4. They are all mammals.

5. (a) Insects don't have backbones.
   (b) The arthropods

6. Porifera

7. The chameleon can blend in with its surroundings.

8. The skunk can emit a very foul odor.

9. Its spines

10. They produce so many seedlings that a few of them do not get eaten and hence grow into mature trees.

11. It feeds on the berries of the plant as well as on insects that have first fed on the plant.

12. Fish rely on algae for their main food source just as cows rely on grass for their major source of food.

13. Deer, birds, mice, rabbits (insects also)

14. The fox

15. They provide food for other organisms.

16. The bacteria and fungi are responsible for causing dead

organisms to decay back into the soil.

17. Some of the rabbits are killed by the parasites.

18. It eats parasites living on the rhino's back.

19. They hibernate.

20. It would probably kill off the young oak and hickory seedlings.

21. The bulldozer destroys their habitat.

22. The English sparrow fed on insects.

23. Such an action would let the fish population once again increase its numbers.

24. (a) Dog breeders would have looked for individual dogs with small stature and long snouts.
    (b) The dachshund is such a dog.

25. It could better search for food.

26. The brain

27. One way is that individuals can warn the entire group of impending danger.

28. (a) Muscular
    (b) Yes

29. The skull gets larger.

30. Large horses could run faster and therefore elude their predators.

31. As North America moved northward, its climate became colder than when it was located near the equator.

32. North America and Africa migrated apart.

33. Farmers can plant the corn seeds themselves.

34. They all contain a nucleus, cytoplasm, and a cell membrane.

35. Step 6

36. They will be exact replicas of the original molecule.

37. (a) 23
    (b) 46

39. A seed

39. A tadpole

40. These crops would prolong the growing season and thus produce more. They would also allow the farmer to plant in areas not previously used because they were too cold.

Chapter 7

METEOROLOGY

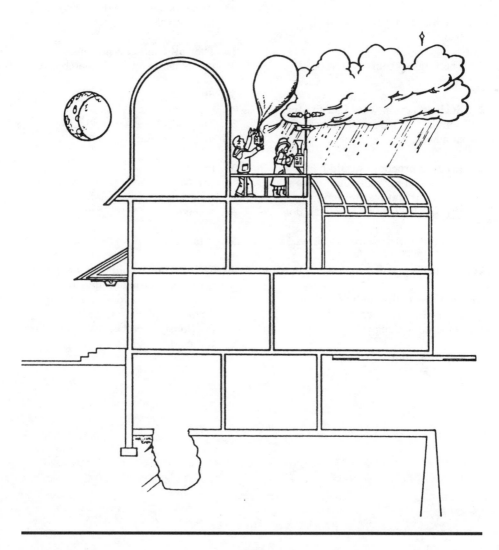

Meteorology is the study of our atmosphere and weather. This science is appropriately located on the roof of the House of Science. From this vantage point, meteorologists are able to collect information pertaining to such things as wind, temperature, and rainfall. They can then put this information together to explain what our weather is likely to be in the future. They can also chart long-term changes in our atmosphere, such as how air pollution may be affecting it.

ACTIVITY

Set up your own weather station with a thermometer, weather vane (or wind sock), and rain gauge, (A wide-mouth tin can will work for this.) After you have recorded the weather for a period of days, give your own weather report!

THE WATER CYCLE

To understand meteorology, you should first know something about the water on our planet, for this water greatly affects our daily weather. You should know, for example, that the earth's water is caught up in a giant cycle of evaporation, condensation, and precipitation. This is known as the *water cycle* (Figure 7.1).

In the water cycle, most of the moisture that evaporates does so from the oceans, simply because they cover so much of the planet.

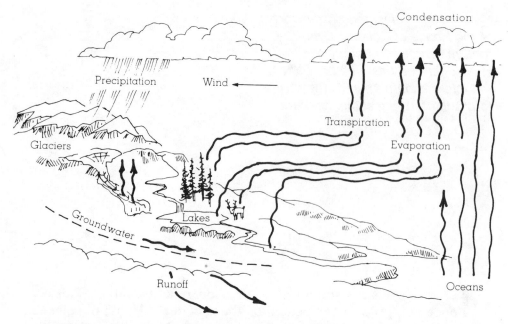

FIGURE 7.1   The Water Cycle (Reprinted by permission from *Meteorology Today* by C. Donald Ahrens. Copyright © 1982 by West Publishing Company. All rights reserved.)

Evaporation can occur from lakes and streams as well, or simply from wet ground. Water is also given off by plants such as the trees shown in Figure 7.1. Water given off in this way is referred to as *transpiration*.

1. When water condenses in the atmosphere (but before the atmosphere gives it back as precipitation), what forms in the sky?

Much precipitation is in the form of rain. Some, however, is in other forms, such as snow, sleet, or hail. In some areas, successive snowfalls build up to eventually form a glacier. Water tied up as glacial ice may stay there for hundreds or thousands of years before finding its way back into the water cycle.

2. Ground water is water that has soaked into the ground following a rainfall or snow melt. In coastal areas, this water may simply seep back into the ocean to begin the water cycle once again. Name another way that water returns to the ocean, thus completing the water cycle.

COMPOSITION OF OUR ATMOSPHERE

Our atmosphere is made up of the air we breathe. The composition of this air is shown in Figure 7.2. Most air varies in composition between dry and humid. The amount of water vapor in the air at any one time is referred to as its *humidity*. Quite simply, the more water vapor in the air, the more humid it is.

Keeping humidity in mind, it is known that warm air holds more moisture than cold air. This is most noticeable on a hot summer day, for example, when the air may feel muggy. This mugginess is due to the great amount of moisture (water vapor) in the air. If this humid air cools, clouds form and rainfall results. This is because the cooler air cannot hold as much moisture.

3. (a) On muggy summer days, thunderstorms often form late in the afternoon, when the sun begins to lose some of its intensity. What usually happens to the air temperature as the sun goes down?

(b) Why can this cause a thunderstorm?

Our atmosphere may be thought of as a "blanket" of air that encircles the planet. In addition to holding moisture, this blanket shields us from harmful types of solar radiation. It also acts as a true blanket by holding in the solar rays that *do* reach the earth's surface (Figure 7.3).

Notice in Figure 7.3 that the earth's atmosphere is thicker over the equator than it is at the poles. This is because it is warmer at the

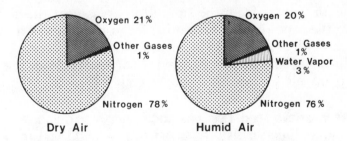

FIGURE 7.2 The Composition of Air (From *Earth Science* by Samuel N. Namowitz and Donald B. Stone. Copyright © 1960 by D. Van Nostrand Company, Inc.)

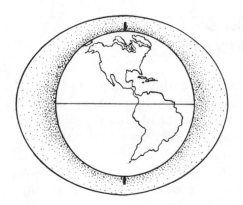

FIGURE 7.3   The Earth's Atmosphere
(*Note:* The thickness of our atmosphere is greatly exaggerated in this diagram.)

equator and warm air expands. Notice also that, as you go upward in the atmosphere, the number of air molecules decreases. This means that the air aloft becomes thinner. You notice this, for example, when your ears "pop" as you drive up a mountain road or ascend in an airplane.

4. Commercial jets generally fly at high altitudes where there is little air resistance to slow them down. On the other hand, there is too little air for people to breathe at these altitudes. What is done to the cabins of these aircraft to ensure that there will be an adequate supply of air for the passengers?

CIRCULATION OF OUR ATMOSPHERE

Most of the direct rays from the sun fall in areas near the equator. As the air near the equator is heated, it begins to rise. As shown in Figure 7.4, it rises until it is very high in the atmosphere, then it divides. When it divides, some of the air travels northward from the equator and the rest travels southward from the equator.

5. As you have seen, rising air cools and, in so doing, is not able to hold as much moisture. Given this, why do you think regions near the equator experience almost daily rainfall?

At about 30 degrees north latitude, air that has been traveling north from the equator has cooled significantly and begins to sink back toward the earth's surface (point A, Figure 7.5). When it does

134

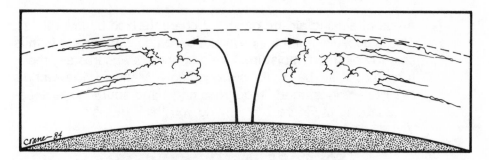

FIGURE 7.4   Air Movement at the Equator

reach the surface, it splits, with a portion traveling back toward the equator and the remainder moving northward.

Figure 7.5 illustrates that air traveling northward from point A eventually encounters cold polar air near point B. This collision of warm, northward-moving air with cold polar air results in many of the storms we experience. It also results in a general rising of air in the vicinity of point B. This rising air at point B in turn splits high in the atmosphere, with a portion of the air traveling northward toward the pole and the remainder traveling southward toward the equator.

When we look at the general circulation of air in the northern hemisphere, we can see that it consists of three revolving "cells" (Figure 7.5). In like fashion, the circulation of air in the southern hemisphere consists of three cells. As you can see in Figure 7.5, each of these six cells of air revolves in the opposite direction of its immediate neighbor.

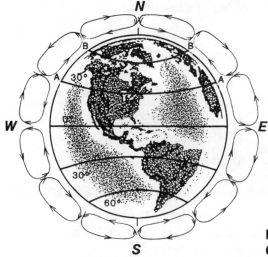

FIGURE 7.5   General Circulation of the Atmosphere

135

6. Areas of sinking air are generally characterized by low rainfall. This is because, as the air sinks, it becomes warmer and is thus able to hold more moisture. Like a sponge, it absorbs any moisture that may be in its surroundings. Many of the world's deserts are located at 30 degrees north and south of the equator. Looking at Figure 7.5, why do you think this is?

FACTORS INFLUENCING OUR WEATHER

Although the overall circulation of the atmosphere plays a role in determining our weather, many other factors go into the making of this weather also. Let's look at two of these factors—the presence of mountains and the presence of nearby oceans.

Mountains

Mountains can have a profound effect on the weather in a region. As shown in Figure 7.6, areas on one side of a mountain can receive plentiful rainfall, while areas on the other side receive very little.

7. In the United States, many areas west of the Rocky Mountains have a humid climate characterized by westerly winds and plentiful rainfall. Areas just east of the Rockies, however, have a desertlike climate. Referring to Figure 7.6, what do you think accounts for these differences in climates?

FIGURE 7.6 How Mountains Affect Weather (From *The Physical World, Second Edition* by Richard Brinckerhoff. Copyright © 1963 by Harcourt Brace Jovanovich, Inc. Reprinted by permission of the publisher.)

8. **(a)** Mountains also affect the temperature in an area. As you go higher in the mountains, what generally happens to the temperature?

(b) Some areas of land near the equator are covered by snow. In what type of terrain are they found?

Oceans

In some areas of the world, the weather is greatly influenced by a nearby ocean. The weather of the British Isles, for example, is greatly influenced by a warm current that flows in the Atlantic Ocean. This current is known as the *Gulf Stream* (Figure 7.7).

9. The warm water of the Gulf Stream heats the air directly above it. It also supplies it with moisture. In wintertime, as this warm, humid air passes over the cold land of the British Isles, it is cooled down. When this happens, the air can no longer hold all of its moisture. As a result, clouds form at ground level. A cloud at ground level is referred to as *fog*. Which city in England is particularly known for its heavy fogs?

The presence of a nearby ocean tends to moderate the weather in an area. This is because the ocean warms up and cools down much slower than the land. You notice this on a hot day, when the land gets unbearably hot yet the nearby ocean stays relatively cool.

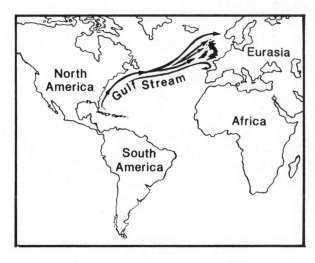

FIGURE 7.7
The Gulf Stream

ACTIVITY

Put a container of sand and a similar container of water side by side in the sun or under a heat lamp. Record the temperature in both containers every minute. Which heats up faster, the sand (representing land) or the water (representing ocean)? Which cools down faster?

OUR CHANGING WEATHER

As air "sits" for any length of time over an area, it acquires the moisture and temperature characteristics of that area. Air positioned over northern Canada, for example, becomes cold and dry like the land under it. On the other hand, air positioned over the Gulf of Mexico becomes warm and humid. In North America, much of our weather is affected by such air masses as they migrate from their source region (Figure 7.8).

10. Looking at Figure 7.8, which two air masses originate over land areas and could therefore be expected to contain relatively dry air?

11. Tropical Pacific air generally holds much more moisture than polar Pacific air, yet both air masses originate over water. What do you think accounts for the difference in moisture content?

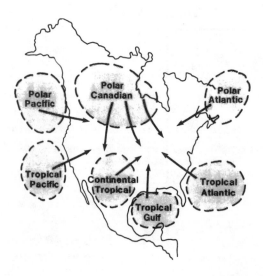

FIGURE 7.8   Air Masses

As air masses move from their place of origin, they greatly affect the weather in their path. Polar Canadian air, for example, brings cool weather in summer and bitter cold weather in winter. On the other hand, tropical gulf air brings hot muggy weather in summer and warm rainy weather in winter.

As air masses move, it is not uncommon for one mass to encounter another (Figure 7.9). Tropical Pacific air may encounter polar Canadian air, for example. When this happens, a *front* is formed.

When warm and cold air masses encounter one another, the cold air, being denser, begins to slide under the warm air. When this happens, the warm air is pushed to higher altitudes where the temperature is cooler. As this warm air cools, it can no longer hold all of its moisture. As a result, clouds form and there is precipitation. This is why stormy weather is often associated with fronts.

12. As you can see in Figure 7.9, a cold front is pushing from the northwest into Denver and surrounding areas. Also, a stationary front exists between warm and cold air in the northern United States up into Canada. What other front is present on this map and over which major city has it just passed?

FUTURE OF OUR ATMOSPHERE AND WEATHER

With the aid of weather satellites, meteorologists are able to closely monitor our atmosphere and its weather. Among their findings is the observation that the ozone layer in our upper atmosphere is being depleted. This depletion is thought to be caused, in part, by certain chemicals our society generates, such as the chlorofluorocarbons (CFCs) used in insulation, styrofoam packaging, refrigerators, and air conditioners. The seriousness of this depletion should not be underestimated, for it is the ozone layer that shields us from the ultraviolet rays of the sun. These ultraviolet rays, you may remember, are what give us a sunburn.

13. Many doctors feel that there will be an increase in skin cancers in the future, due in part to people getting sunburned more often. What does this have to do with the ozone layer in our atmosphere?

139

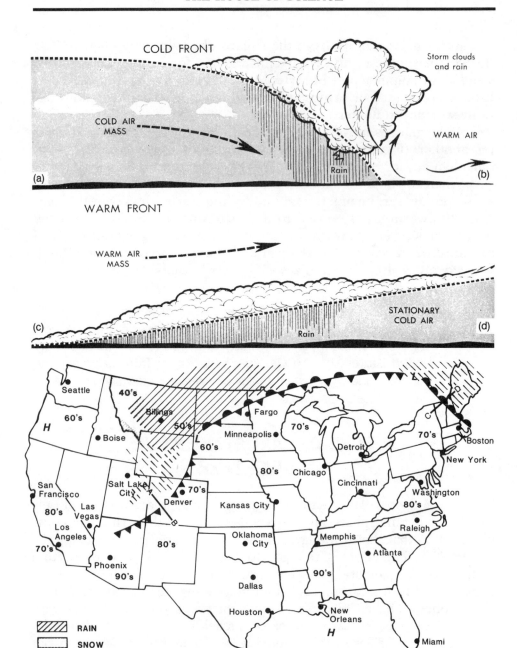

FIGURE 7.9 Weather Fronts (Warm and cold front diagrams: From *The Physical World, Second Edition* by Richard Brinckerhoff. Copyright © 1963 by Harcourt Brace Jovanovich, Inc. Reprinted by permission of the publisher.)

Meteorologists also have discovered that there is more carbon dioxide in our atmosphere than there used to be. This increase in CO_2 is thought to be caused by the burning of fossil fuels such as coal and oil. It is also due to the widespread destruction of forest areas, particularly tropical rain forests.

Carbon dioxide is effective in trapping heat that is trying to leave our atmosphere. As a result, this gas helps to keep our atmosphere warm. It is believed that too much carbon dioxide in the atmosphere, along with other waste gases, will lead to a general global warming known as the *greenhouse effect* (Figure 7.10).

The greenhouse effect is beginning to change climates in some areas, along with the types of crops that can (or cannot) be grown in those areas. Also because of the greenhouse effect, we are beginning to see a rise in sea levels around the world. This rise is caused by the melting of glacial ice.

14. Hurricanes form over warm ocean waters. They in turn are known to increase in intensity as they pass over warm waters. If the greenhouse effect continues as predicted, why will hurricanes of the future probably be worse than hurricanes of the past?

SUMMARY

Meteorology is appropriately located on the roof of the House of Science. From this vantage point, scientists can collect data pertaining to such things as wind, temperature, and rainfall. With this

in a greenhouse

in a car

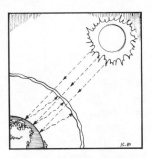
in the Earth's atmosphere

FIGURE 7.10 The Greenhouse Effect

information, they can better understand our atmosphere and the weather we experience.

To understand meteorology, you should first know something about the water on our planet. This water is caught up in a giant cycle of evaporation, condensation, and precipitation. This cycle, known as the *water cycle*, greatly affects the weather we experience.

Our atmosphere is made up mainly of nitrogen and oxygen, with small (but very important) amounts of water vapor. This atmosphere acts both as a shield against harmful radiation and as a blanket for holding in solar rays that *do* reach the earth's surface.

In each hemisphere, the circulation of the atmosphere may be thought of as occurring in three large revolving cells. Although this general circulation affects the weather, the weather in any one region is influenced by other factors as well, such as mountains and nearby oceans.

Much of the changing weather we experience is due to the presence of air masses. As these air masses migrate from their source area, they bring with them the temperature and humidity characteristics of that area. And, as they encounter other masses of air, fronts are formed and precipitation may occur.

Our atmosphere and weather will likely be influenced in the future by such things as ozone depletion and the greenhouse effect.

15. Weather forecasting became much more accurate when the telegraph was invented. Why do you think this was?

ANSWERS

1. Clouds

2. Runoff from rivers and streams

3. **(a)** The air temperature usually drops.
   (b) As the air temperature drops, not as much moisture can be held by the air. Therefore, it gives up some as precipitation.

4. The aircraft cabins are pressurized.

5. As shown in Figure 7.4, areas near the equator are characterized by rising air.

6. Areas 30 degrees north and south of the equator are characterized by sinking air.

7. Air rising over the mountains cools, causing it to drop its moisture west of the Rockies. As this now dry air descends over the mountains, it becomes warmer and, therefore, even dryer. This lack of moisture accounts for the deserts just east of the Rockies.

8. (a) It decreases
   (b) Mountainous terrain (high altitudes)

9. London

10. Continental tropical and polar Canadian

11. The Tropical Pacific air is warmer, and warmer air can hold more moisture.

12. A warm front just passed over the city of Boston (moving northeast into Maine).

13. Less ozone would let more damaging ultraviolet light in. This ultraviolet light is responsible for sunburns, which may lead to skin cancer.

14. Hurricanes of the future will probably have more warm water to travel across and thereby be "fueled."

15. The weather conditions from one area of the country could quickly be broadcast to another area, so that people could be notified of impending storms, and so on.

Chapter 8

OCEANOGRAPHY

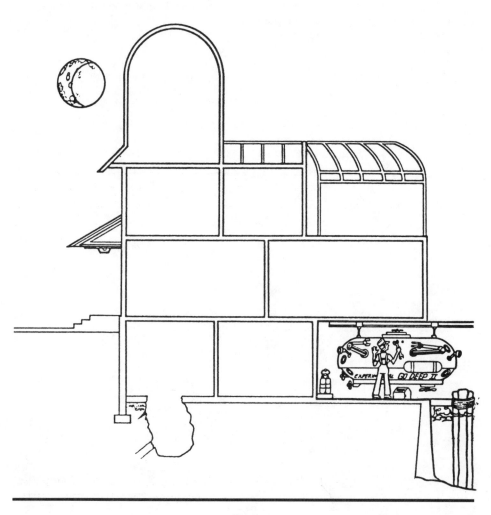

In the House of Science, oceanography is studied using a diving craft. Scientists collect information from the oceans and bring it back to the house for analysis. Oceanography includes a study of the ocean floor, currents, and tides, as well as the plants and animals that live in the oceans.

When looking at a world map, you notice that our earth is covered more by oceans than by land. In fact, a little over two-thirds of our planet is covered by ocean areas. These oceans are used for many purposes, such as fishing, transportation, and recreation. But they are important in other ways too. Ocean currents, for example, transport warm water away from the equator toward the poles, thereby warming those areas. Other currents return cold water to the equator to begin the cycle again. In this way, much of the solar heat our earth receives is distributed around the planet. In addition to distributing heat, the oceans also serve as home for much of the life on our planet. Without this ocean life, our earth would be a quite different world.

Oceanography is the study of the world's oceans. In the House of Science, oceanography is studied using a diving craft. As you can see, this craft is being readied for an expedition at sea. Information collected from this expedition will be brought back to the house and analyzed. With this information, we may come to know more about the vast potential of our oceans.

TOPOGRAPHY OF THE OCEAN FLOOR

At one time, people thought that the bottom of the sea was pretty much a flat, featureless plain. This was because detailed maps of the ocean bottom did not exist. The little knowledge we did have of the ocean floor was from depth readings taken by lowering a line over the side of a ship. This method was, needles to say, very tedious and time-consuming. As a result, not many readings were taken.

In the 1920s a device called an *echo sounder* was invented. As seen in Figure 8.1, this device operates by bouncing sound waves off the ocean floor. The length of time it takes for the sound wave to re-turn is noted, and the depth to the ocean floor can be calculated.

1. At location A in Figure 8.1, the sound wave took one second to return to the ship, indicating that the depth of ocean bottom was about one-half mile (.8 km). At location B, the sound wave took three seconds to return to the ship. What was the depth of the ocean floor at location B?

The invention of the *echo sounder* (also called *sonar*) gave us "eyes" to look at the floor of the ocean. With this device, a ship can now take thousands of depth readings on any one voyage. This is be-cause it is able to keep moving as it is taking the readings, unlike the old method where the ship first had to be stopped before a line could be lowered over the side. Figure 8.2 shows a picture of the ocean floor that emerged as a result of sonar readings.

2. **(a)** Does the floor of the Atlantic Ocean appear to be a flat, featureless plain?

(b) What feature runs through the center of the Atlantic Ocean?

Continental Shelves

Continental shelves are relatively shallow, gently sloping areas found adjacent to continents. In Figure 8.3 they appear as shelflike

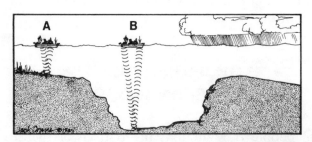

FIGURE 8.1 Mapping the Ocean Bottom by Echo Sounding

146

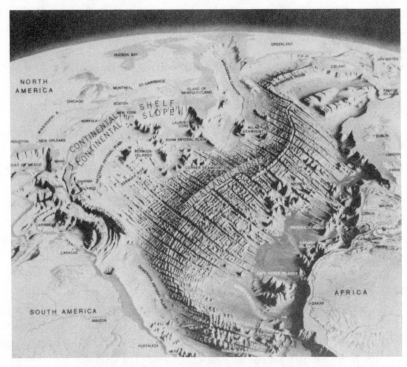

FIGURE 8.2   View of the Atlantic Ocean Floor if All the Water Were Removed (Courtesy of Aluminum Company of America)

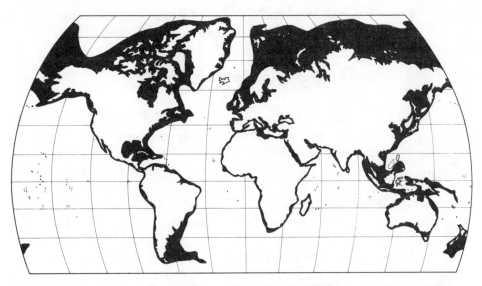

FIGURE 8.3   Continental Shelves (From *Physical Geology* by Richard Foster Flint and Brian J. Skinner. Copyright © 1974, 1977, by John Wiley & Sons, Inc.)

extensions of the continents. Indeed, they may be thought of as parts of the continents that just happen to be under water.

3. Do the continental shelves appear to be of about equal width in all places of the world?

4. Life is not evenly distributed throughout the oceans. Rather, 90 percent of ocean life is found on the continental shelves. Ironically, most ocean pollution also occurs in these continental shelf waters. Name one way that we pollute the water adjacent to continents.

There were times in the past when large areas of the continental shelves were exposed above sea level. This happened during past ice ages when much water that *had* been in the oceans was now tied up as ice. It was contained in ice sheets similar to those found today over Antarctica and much of Greenland.

Today, coastal cities the world over would be flooded if the ice on Antarctica and Greenland were to melt significantly. This is because water that was tied up as ice would now be in the oceans, thereby raising the level of the sea and flooding coastal areas.

ACTIVITY

From a world map, make a list of major coastal cities of the world that would be affected by a rise in sea level.

CHEMISTRY OF SEA WATER

Sea water is composed of 96.5 percent water and 3.5 percent dissolved elements (Figure 8.4). As shown, the two major elements dissolved in the sea are sodium and chlorine. These two elements combine to form sodium chloride—salt. This is what gives sea water its characteristic salty taste.

Among those elements listed as "others" in Figure 8.4 are rare elements such as gold. It is estimated that there are millions of tons of gold dissolved in sea water. Removing this element would not be cost-effective, however. That is, it would cost more than the gold was worth.

5. If all the gold dissolved in the oceans could be removed, gold would cease to be a valuable commodity. Why?

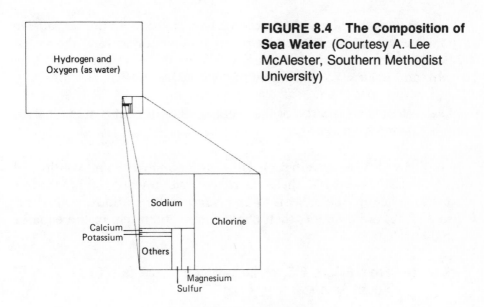

FIGURE 8.4 The Composition of Sea Water (Courtesy A. Lee McAlester, Southern Methodist University)

OCEAN CURRENTS

Most of the water in the oceans is in constant motion. At the surface, this motion is determined largely by global wind patterns and by the position of the continents (Figure 8.5).

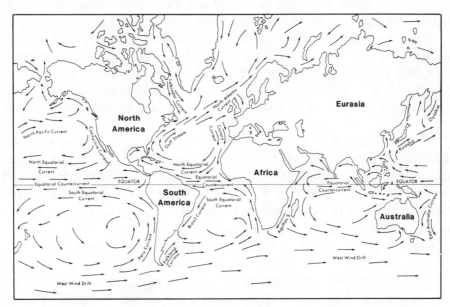

FIGURE 8.5 Major Surface Currents (U.S. Navy Hydrographic Office)

Along with the atmosphere, the surface currents in the oceans are responsible for transferring heat from the equator toward the poles. Without this heat transfer, tropical areas would be much hotter than they are and polar areas would be much colder.

6. What is the source of the great amount of heat that is received near the equator?

In Figure 8.5, warm currents flow both northward and southward away from the equator. As these currents travel toward the poles, they gradually give up their heat to the surrounding atmosphere, thus warming it. These currents then begin their return trip to the equator as cold currents.

7. (a) From Figure 8.5, name two cold currents off the coast of South America.

(b) The British Isles are warmed by a current that also flows along the eastern shore of the United States. What is the name of this ocean current?

In addition to surface currents, there are deep currents in the oceans. Many of these currents originate in polar areas where the water becomes quite cold and dense. As a result, this water sinks to the bottom and flows toward the equator as bottom currents.

Other currents arise because of differences in the salt content (salinity) of the water. Water in the Mediterranean Sea, for example, has a higher salinity than normal ocean water. This is due to evaporation in the dry air surrounding the Mediterranean. Because of this salinity difference, water from the Mediterranean Sea weighs more than normal ocean water.

Figure 8.6 shows that, at the Straits of Gibraltar, dense water from the Mediterranean Sea meets less dense water from the Atlantic Ocean. The Mediterranean water, being heavier, flows into the Atlantic as a bottom current. This water is in turn replaced by water from the Atlantic flowing into the Mediterranean as a top current.

8. In World War II, Allied ships were stationed in the Straits of Gibraltar. They were there to listen for the engines of German submarines that may have been trying to enter the Mediterranean Sea. Some German subs were able to pass both ways through the straits, however, without being detected (Figure

150

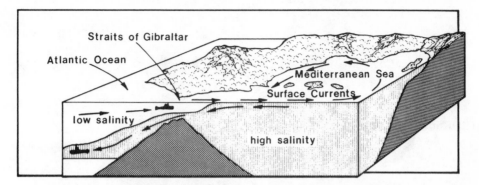

FIGURE 8.6 Flow of Water through the Straits of Gibraltar (From *Introduction to Geology: Physical and Historical,* by William Lee Stokes and Sheldon Judson © 1968, pp. 223. Reprinted by permission of Prentice Hall, Inc., Englewood Cliffs, New Jersey.)

8.6). Why were the Allied ships unable to detect the sounds of their engines?

9. As mentioned, the high salinity of the Mediterranean Sea is due to the warm dry climate in that area. As water evaporates in this enclosed area, it serves to concentrate the dissolved elements already present in the sea water. When water evaporates, does it take the dissolved elements along with it or does it leave them behind?

TIDES

Tides are a result of the pull of gravity from both the moon and the sun. The moon's gravitational pull is greater than the sun's because the moon is much closer to us.

As shown in Figure 8.7, the moon creates a bulge in the ocean waters on the side of the earth that faces it. The moon also tends to pull the earth "out from under" the opposite side, thereby creating a bulge there also.

As Figure 8.7 shows, when the sun and moon are pulling on the earth at the same time, the result is higher than normal tides called *spring tides.* (*Note:* Spring tides have nothing to do with the season of year. That is just the name they were given.) Figure 8.7 also shows that, when the sun and moon are pulling at right angles to each other, we have lower than normal tides called *neap tides.*

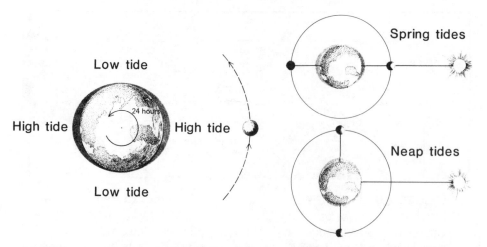

FIGURE 8.7 Tides. Left: The effect of the moon in creating tides; right: The effect of the sun and moon in creating tides. (Courtesy A. Lee McAlester, Southern Methodist University)

10. The moon's gravity pulls on land areas as well as areas of water. However, bulges don't normally show on land as they do on water. Why not?

The bulge of the ocean waters shown in Figure 8.7 is greatly exaggerated. In reality, the oceans bulge out less than three feet. As the earth rotates, however, this tidal bulge can become "funneled" into river inlets. If the shape of the inlet is right, the height of the tide can be greatly magnified. This is the case in the Bay of Fundy in Nova Scotia, Canada, where the tidal range can be as great as 50 feet (15 meters) (Figure 8.8)!

11. In the Bay of Fundy, why is it important for fishermen to know when the tide is going to be low?

WAVES

Waves are generated by the wind. The strength and duration of the wind affects the size of waves, as does the area over which these winds blow. Larger waves can be generated in the Pacific Ocean, for example, than in the Great Lakes because the wind has a much greater area over which to work.

FIGURE 8.8   High Tide and Low Tide in the Bay of Fundy

When waves reach the shoreline they erode and sculpture it (Figure 8.9). Many factors are involved in how the shoreline will look; thus, a whole branch of oceanography is devoted to the study of shorelines.

12. Figure 8.9 shows the profile of a shoreline that has been subjected to the pounding of waves. Where do you think the material for the wave-built terrace came from?

LIFE IN THE OCEANS

Just as on land, there exists a food chain or "web of life" in the ocean as well. Microscopic plants such as algae serve as the base for much of this life. In all, these plants make up 90 percent of the weight (called *biomass*) of all life in the oceans. These plants (*phytoplankton*) in turn serve as food for microscopic animals (*zooplankton*). Both microscopic plants and animals then serve as food for larger animals in the ocean (Figure 8.10). Some of these larger animals, such as fish, are in turn fed upon by still larger animals, such as sharks.

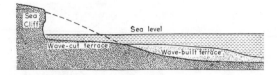

FIGURE 8.9   The Formation of a Shoreline (Schematic: From *Earth Science* by Samuel N. Namowitz and Donald B. Stone. Copyright © 1960, by D. Van Nostrand Company, Inc. Photograph: Photo Researchers, Inc.)

13. It is ironic that the largest animal in the sea feeds almost entirely on the smallest life forms—phytoplankton and zooplankton. Looking at Figure 8.10, what is the largest animal found in the sea?

ACTIVITY

Compile your own "Who's Who" picture catalog of sea life. Entitle it *Denizens of the Deep.*

When organisms living in the ocean die, their remains sink toward the ocean bottom. Thus, the lower ocean layers receive a constant "rain" of nutrients from above. In certain areas of the world, such as off the coast of Peru, these lower ocean layers are sometimes forced to the surface by currents. This is shown by the large arrow in the left of Figure 8.10.

When nutrients are forced to the surface, they serve as food for the phytoplankton, causing these microscopic plants to multiply rapidly. This "bloom" of phytoplankton in turn allows other organisms, such as fish, to increase their numbers. In some areas, therefore, this upwelling of waters is very important to the fishing industry.

14. In the future, how might pumping water from the ocean depths to the surface allow us to operate a fish "farm"?

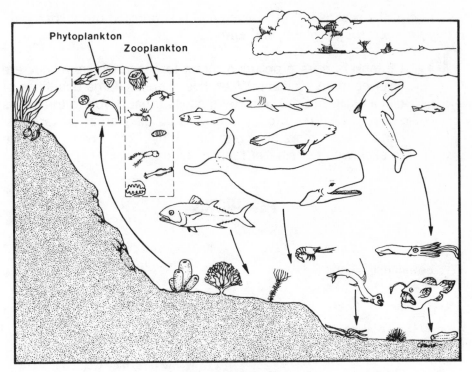

FIGURE 8.10   Ocean Life

SUMMARY

The oceans of our planet cover a little over two-thirds of the globe. We study these oceans today with vessels such as the diving craft shown in the House of Science. What we find from such studies is that, in places, the ocean floor is as rugged as many land areas. Among the features of this ocean floor are the continental shelves directly adjacent to continents. These gently sloping areas contain 90 percent of the life found in the sea. In times past, during ice ages, parts of these continental shelves were exposed as land.

Ocean water is made up of 96.5 percent water and 3.5 percent dissolved elements. Much of this water is in constant motion due to currents. Surface currents are determined largely by global wind patterns, along with the position of continents. Currents beneath the surface are driven mainly by differences in water density.

Ocean tides are caused by the gravitational pull of the moon and sun, particularly the moon. Ocean waves, which are caused by the wind, often help to "sculpture" a shoreline.

Life in the oceans is quite varied and is supported directly or indirectly by microscopic plants and animals.

15.
Oil tankers pose a pollution threat to ocean waters if they are involved in an accident. When this happens, large quantities of oil can spill into the water and cause great havoc to the area's ecology. The trend today is toward building larger tankers that can transport more oil at one time. Why might it be a good idea to stick with the smaller tankers instead?

ANSWERS

1. $\dfrac{1 \text{ sec.}}{^1/_2 \text{ mile}} = \dfrac{3 \text{ sec.}}{x}$; now
cross-multiply:
$1x = 3 \times {}^1/_2 \text{ mile}$
$x = \dfrac{3}{2} \text{ miles}$
$x = 1^1/_2 \text{ miles deep}$

In metric:

$\dfrac{1 \text{ sec}}{.8 \text{ km}} = \dfrac{3 \text{ sec}}{x}$; now cross-multiply:
$1x = 3 \times .8 \text{ km}$
$x = 2.4 \text{ km deep}$

2. **(a)** No
   (b) The Mid-Atlantic Ridge

3. No. They vary widely.

4. Ocean dumping, oil spills, and so on.

5. There would be so much of it that it would no longer be scarce and, therefore, valuable.

6. Sunlight

7. **(a)** The Peru Current and the Falkland Current
   (b) The Gulf Stream

8. These subs "cut" their engines and drifted through on the currents.

9. It leaves them behind in the remaining water.

10. Unlike water, land is rigid.

11. At low tide, they would be stranded on land without being able to go to sea. If at sea, they would not be able to return to port at low tide.

12. The wave-cut terrace

13. The whale

14. By pumping water from the depths, we can bring nutrients to the surface. These nutrients can in turn be used to feed microscopic plants and animals—the food for fish.

15. If a smaller tanker has an accident, smaller amounts of oil are released to the ecosystem. (Also, smaller tankers are more maneuverable and, therefore, less likely to get in an accident.)

Chapter 9

ENVIRONMENTAL SCIENCE

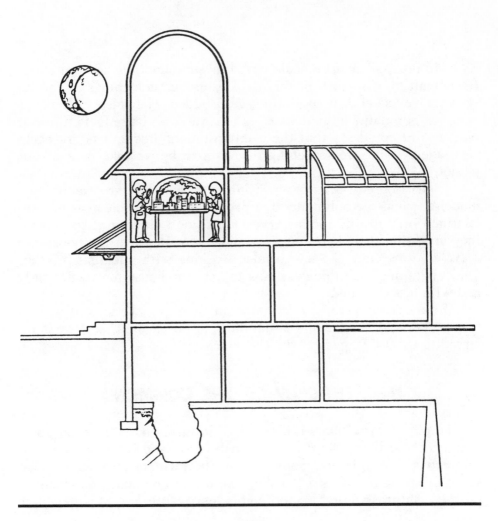

Environmental scientists study the resources we produce and use, the pollution we create, and the wastes we dispose of. Environmental scientists work with scientists in other fields to improve energy efficiency and fight pollution. These scientists make sure that the residents of the house recycle many of the items they use.

E nvironmental science deals with our surroundings here on earth. Scientists who work in this field examine such things as the resources we use and the pollution we produce. They also look at our current population "explosion" and evaluate its effects on our planet. You could say that these environmental scientists are earth "managers" who are searching for ways to better take care of our planet.

In the House of Science, environmental scientists are examining a model city and the pollutants it is producing. These scientists are not confined only to this room, however. As good citizens of the house, they are using their knowledge to make sure items in the house are recycled, for example. They are also working with scientists in other fields, pursuing such things as how to heat the house more efficiently and with less pollution.

To better understand environmental science, let's look at a story that pertains to our current situation here on earth. This story is referred to as *The Tragedy of the Commons*.

THE TRAGEDY OF THE COMMONS

The *Tragedy of the Commons* takes place in England. As the story goes, a group of farmers shared a common field of land referred to as the "commons." On this commons, one of the farmers decided to raise some sheep. To his delight, the sheep he raised grew large and healthy on the lush grass found on the commons. Before long, his neighbor took note of this and decided to raise sheep there also. These sheep ate well and grew large too. As a result, it was not long before other farmers "discovered" the commons and decided to graze sheep there as well. But an interesting thing happened as more sheep grazed the land. Instead of the animals becoming large and healthy, they all became weak and sick. This was because there was not enough grass in the field to support all of them. As a result, none of the sheep fared well. What was good for a few, therefore, was not good for any as the commons became overpopulated (Figure 9.1).

The story of *The Tragedy of the Commons* is much like the environmental crisis we face today on earth. As our numbers increase, the competition for available food likewise increases, so that today millions of people are malnourished or even dying of starvation. In addition, as our population increases, the demand for natural resources also increases. As supplies of these resources dwindle, we are forced to use alternative sources, or in some cases, to do without. Finally, as our population increases, the amount of wastes we generate likewise increases. In many cases, these wastes are harmful to our overall

FIGURE 9.1 The Tragedy of the Commons

health and well-being. Figure 9.2 depicts the relationship between population, resource use, and the wastes we generate.

1. As population increases, what happens to the amount of resources we use and the amount of wastes we generate?

2. Some food shortages are caused not by the lack of food, but by a failure to distribute it to areas where it is needed. Fresh fruit grown in abundance in one area of the world, for example, cannot always be transported to where it is needed. Why might this be?

To understand environmental science, let's look separately at each of the points of the environmental triangle shown in Figure 9.2—population, resource use, and waste disposal. Let's use these three points as guideposts to understanding our environment.

POPULATION

A picture of the world's population can best be seen by looking at a graph of population versus date (Figure 9.3).

3. Notice on Figure 9.3 that the world's population held steady at about 500 million people for much of recorded history. By about 1800, however, the population began to climb, so that by about 1850 the population had doubled to about one billion people. At this point, population growth began in earnest. What was the world's population at each of these dates: 1900, 1930, 1960, 1976, 1990?

FIGURE 9.2   The Environmental Triangle

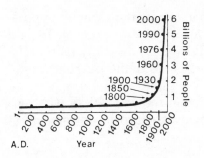

FIGURE 9.3 Population Growth (From *Living in the Environment*, Second Edition, by G. Tyler Miller, Jr. © 1979 by Wadsworth, Inc. Reprinted by permission of the publisher.)

4. Why do you think population growth in recent years has been referred to as the population "explosion"?

Population Growth

To illustrate how population grows, let's look at two countries with similar populations, but quite different growth rates: Great Britain and Egypt (Table 9.1).

5. **(a)** Referring to Table 9.1, Egypt will have how many more people than Great Britain in the year 2000?

(b) How many more people will Egypt have than Great Britain in the year 2025?

Egypt currently has a growth rate of 2.4 percent, which means that there will be 2.4 percent more people living in that country next year than there are this year. In contrast, Great Britain has a growth rate of only 0.04 percent. This difference in growth rates means that Egypt's population is growing 60 times faster than that of Great Britain!

TABLE 9.1 Population Growth in Egypt and Great Britain

Year	Egypt (Growth Rate: 2.4%/Yr.)	Great Britain (Growth Rate: 0.04%/Yr.)
1990	52.7 million	55.8 million
1995	58.9 million	56.0 million
2000	65.2 million	56.2 million
2025	97.4 million	56.4 million

Source: Population Division of the United Nations

To find Egypt's population for any given year, multiply its population the previous year by 1.024 (its growth rate of 2.4 percent). To find the population of Great Britain for any given year, multiply its population the previous year by 1.0004 (its growth rate of .04 percent).

6. If Egypt's population is 65.2 million in the year 2000, what will its population be in 2001, assuming its growth rate remains at 2.4 percent?

7. At its current rate of growth, by the year 2029 Egypt will have doubled its population from what it was in 1990. How many years will it have taken for this doubling to occur?

Although a growth in population of 2.4 percent a year may not seem like a lot, you can see from Table 9.1 that it *will* result in large increases in population. Table 9.2 shows populations and projected growth rates for various countries and for the world as a whole.

8. (a) From Table 9.2, list the two countries with the highest growth rates.

(b) Which country has a negative growth rate and is, therefore, actually declining in population?

TABLE 9.2 Populations (in Millions)

	1990	1995	2000	2025	Current Growth Rate/Yr.
Australia	17	18	19	24	1.2%
Brazil	150	165	180	246	2.0%
Canada	27	28	29	34	0.9%
China (mainland)	1,120	1,184	1,256	1,460	1.1%
Denmark	5.2	5.1	5.1	4.8	−0.4%
Ethiopia	43	50	58	112	3.4%
India	832	899	962	1,188	1.6%
Kenya	25	31	38	83	4.7%
Mexico	89	99	109	154	2.2%
Soviet Union	291	303	315	367	0.8%
Sweden	8.2	8.2	8.1	7.5	0.0%
United States	248	258	268	313	0.8%
World	5,248	5,679	6,127	8,177	1.6%

Source of population figures: Population Division of the United Nations

9. If the world overall has a growth rate of 1.6 percent, what does this mean when comparing this year's population with next year's?

10. At the present rate of growth, the population of the world increases by about 90 million people each year. Considering that New York City has about 7.1 million people, we are adding the population equivalent of how many New York Cities to the world every year?

Future Population Growth

Given the present growth rate of 1.6 percent, the world's population will more than double in your lifetime. But is this rate of growth likely to remain the same? With a significant portion of the world's population currently malnourished or undernourished, can the earth sustain a population more than twice the size it has now? Some scientists think it can, but a great many think it is unlikely, given our present resources and technology.

We *have* made strides in feeding the world's population. For example, scientists have come up with new strains of wheat, rice, and other crops that produce a good deal more per plant than the old strains did. These new strains are part of what is known as the *Green Revolution*. In some areas, this "revolution" has seen a great increase in farm yields.

Although the Green Revolution has brought us increased crop yields, in many areas of the world population growth has already outstripped the effects of this revolution. The result has been simply a greater number of people than ever before living at or below the subsistence level.

11. In looking at the *Tragedy of the Commons*, we can liken the Green Revolution to the plants in the field growing twice as fast as they normally would. Thus, the field could support twice as many sheep than it did before. How would a constant increase in the number of sheep eventually result in none of them getting enough food?

The world's current growth rate cannot continue indefinitely, because the world obviously cannot support an infinite number of people. Our numbers will be held in check in the future, perhaps by mass starvation or by war. A better alternative is for us to do it voluntarily by bringing our growth rate down to zero. This is what is called *zero population growth* (ZPG). When the world reaches ZPG, the birth rate will match the death rate and our population will no longer increase.

12. A few of the European countries have reached zero population growth and a few are actually declining somewhat in population. If a country had, say, 8.3 million people in 1985, and had attained ZPG, what would be its population in the year 2000?

In several countries, efforts have been made toward curbing population growth. For example, in India and China, the world's two most populous countries, measures have been tried such as mass sterilizations, free dispersal of contraceptives, and financial incentives such as tax breaks for having a small family. In each of these countries, couples have been educated to plan for a small family, rather than letting the laws of nature dictate the size family they will have. China has had somewhat more success than India in curbing its population, although neither country has yet reached ZPG.

13. In 1988, India had a population of about 800 million and China had a population of about 1,100 million (1.1 billion). Of the world's approximately five billion people, what percentage lived in each of these countries in 1988?

Population Distribution

The world is sometimes divided into two categories, based on standard of living. These two categories are the developed nations and the developing nations.

Developed nations are industrialized; these countries include the United States, Canada, the European countries, Russia, Japan, and Australia. They contain 25 percent of the world's population and they use 80 percent of the world's natural resources. Because of their resource use, they are generally high polluters. These countries typically have a low rate of population growth.

Developing countries include India, mainland China, much of Africa, and much of South America. These countries have 75 percent of the world's population but use only 20 percent of the world's natural resources. As a result, they have a quite different standard of living than the developed countries. Malnutrition is common, illiteracy is high, and population growth is rapid. This rapid growth in population means that these countries either have a population problem now or will have one in the very near future.

14. It is estimated that a person in an industrial country uses well over ten times as much energy and natural resources as a person in a developing country. Given this, in what way might a small increase in population in an industrial country be as serious a problem as a large increase in a developing country?

NATURAL RESOURCES

To support our population, various natural resources are needed. Among these resources are the minerals we use in manufacturing and the energy that powers our machines and appliances. Let's take a look at these mineral and energy resources.

Mineral Resources

Mineral resources can be classified into two groups—metals and nonmetals. Metals include elements we are familiar with, such as iron, aluminum, copper, and gold. Nonmetals consist of such things as building stone, rock salt, and the clay used for making bricks. Figure 9.4 shows some of the methods we use for extracting mineral resources.

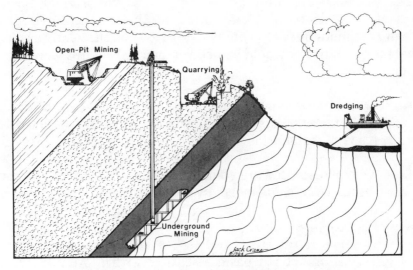

FIGURE 9.4   Methods of Extracting Minerals

15. (a) Open-pit mining is generally cheaper and safer than underground mining. Why do you think this is?

(b) How are deposits on river bottoms recovered?

Some mineral resources are so abundant that it is unlikely we will ever run out of them. Sand and gravel deposits, building stone, and aluminum ore, for example, are all quite abundant. Other mineral deposits, especially those containing metals, are more scarce. New discoveries will be made, but it should be kept in mind that there *is* a finite amount of these resources. They will not last forever.

To avoid running out of mineral resources in the future, we will have to alter our living habits so that we consume less of them. We will also have to practice conservation techniques such as recycling.

16. As some minerals become depleted, we can obtain more supplies by mining lower grade ore. This is the case with copper and iron, for example. Mining lower grade ore requires more energy, however, and is generally more destructive to the environment. This is because of the large amount of waste produced. For some minerals, such as mercury, lower grade ores do not exist. Why might the use of mercury in the future have to be greatly curtailed?

Substitutions can be made for some resources that are growing scarce. The substitute is not always as good as the original, however. For example, in electrical wires aluminum can be substituted for copper. But aluminum does not conduct electricity as efficiently as copper, so that more electricity would be needed than if the wires were made of copper.

17. New cars have considerably less metal than cars of the past. What material has been substituted in parts that used to be made of metal?

18. Mineral resources are sometimes referred to as "one crop only" resources. What is meant by this?

Energy Resources

Today, energy is one of our basic essentials for life. We use it to heat our homes, run our appliances, and power our automobiles. Energy is likewise consumed in making the products we use and in transporting these products to us. Energy is also used in making fertilizers for crops and in powering the machinery to harvest these crops. As you can see, energy is very much a part of our daily lives. Without it we would have a much lower standard of living. Let's look at some of the energy resources being used today and at some sources that may be used in the future.

Oil and Natural Gas. In continental shelf areas, oil and gas are believed to form from the remains of microscopic organisms, mainly plants. These remains drift to the ocean bottom and accumulate with sediment that was brought there from rivers and waves. As this mixture of sediment and organic material becomes compacted under layers deposited on top of it, chemical reactions occur which—over millions of years—turn the organic matter into oil and gas.

Newly formed droplets of oil and gas tend to separate from the water that surrounds them. Eventually they collect together and float to the top of this water. If nothing stops their upward migration, they may eventually escape at the surface as an oil or gas seep. If an impervious rock layer is present, however, the oil and gas may become trapped and accumulate to form a reservoir (Figure 9.5).

19. In Figure 9.5, well A is an exploratory well that shows only a trace of oil. Well B, however, *does* yield oil, and well C yields natural gas. When pumping from well B begins, the oil level will eventually fall below the bottom of the well and no oil will be extractable. Pumping water *into* well A would allow well B to once again pump oil. Why do you think this is?

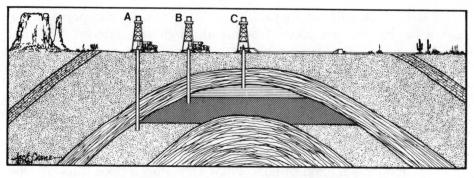

FIGURE 9.5   A Trap for Oil and Natural Gas

Oil and natural gas account for 65 percent of the energy used in the world today. Although these two fuels are widely used, they are not without their problems. For one thing, supplies are dwindling. The search for new supplies has led to deeper wells than ever before and to areas of the world that are remote and inhospitable.

Another problem with oil and natural gas is that they pollute. Oil pollutes a good deal more than natural gas, but both of these fuel sources release carbon dioxide and other pollutants into the atmosphere. Carbon dioxide, you may remember, can increase the greenhouse effect of the earth's atmosphere, thereby leading to an overall warming of the planet.

20. Natural gas is used in cooking, home heating, and in generating electricity. Name two ways that oil is used.

Coal. As seen in Figure 9.6, coal forms in swamp areas. As generation after generation of plants live and die in these swamps, thick deposits of peat build up. With time, the bottom layers of this peat compact under the weight of overlying material. After millions of years of heat and pressure, these layers gradually turn into coal.

21. Coal is mined both at the surface and underground. At the surface, *strip mining* is done by giant shovels that scoop out the coal. Underground mining is done by drilling "rooms" into a coal seam. When mining such a "room," some of the coal cannot be removed. Instead it must be left standing as pillars. Why is this?

Coal accounts for about 30 percent of the energy used in the world today. It was used even more extensively in the past, before oil and gas came into widespread use. As these petroleum supplies dwindle, the use of coal may once again climb.

Coal is not without its problems. Strip mining can permanently scar a landscape, and water flowing through an underground mine can become polluted and make a nearby stream bright orange in color. This polluted water is known as *acid mine drainage*. In addition to these problems, the burning of coal causes problems in the form of air pollution and acid rain.

ACTIVITY

Examine a lump of coal. What appearance does it have? How does it compare in weight to another rock of similar size?

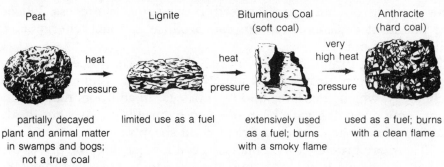

Peat	Lignite	Bituminous Coal (soft coal)	Anthracite (hard coal)
partially decayed plant and animal matter in swamps and bogs; not a true coal	limited use as a fuel	extensively used as a fuel; burns with a smoky flame	used as a fuel; burns with a clean flame

FIGURE 9.6 The Formation of Coal (Top: From *View of the Earth, An Introduction to Geology* by John J. Fagan. Copyright © 1965 by Holt, Rinehart and Winston, Inc. Reprinted by permission of the publisher. Bottom: From *Living in the Environment,* Second Edition, by G. Tyler Miller, Jr. © 1979 by Wadsworth, Inc. Reprinted by permission of the publisher.)

Hydroelectric Power.   Hydroelectric power is used in areas where a dam has been built on a river. Water released near the base of the dam spins turbines, which in turn generate electricity.

There are not many problems associated with hydroelectric power. This source of energy is relatively pollution-free, although some wildlife habitats are destroyed when the reservoir behind the dam first forms. Some dams also prevent the migration of fish, such as salmon, although this problem can be alleviated by building a stepped spillway to one side of the dam. The life of hydroelectric dams is limited to between 50 and 100 years because silt eventually builds up behind them.

About two percent of the world's energy is generated by hydroelectric plants. This percentage is not likely to rise a great deal because dams have already been built on many of the favorable river sites. However, renovation could be done on many small hydroelectric plants that were abandoned in the past. Bringing these small plants back into service would save utility companies the cost of having to build new plants.

22. Why might hydroelectric power have to be curtailed somewhat during times of drought?

Nuclear Power. About two percent of the world's energy needs today are supplied by nuclear power. To generate this power, uranium atoms are bombarded by neutrons and thereby split in half. Tremendous amounts of energy are released in this splitting process. Indeed, one pound of enriched uranium can produce as much energy as 25 railroad cars of coal!

Producing uranium fuel is a fairly inexpensive process. Converting this fuel into electricity is not! Nuclear plants require advanced technology, which translates into high prices. In addition, many of the plants now operating have been plagued by problems. As a result, the building of nuclear plants has dropped off sharply.

There are problems with nuclear power other than the price. One of these problems is the question of where to dispose of radioactive wastes that are generated by these plants. These wastes will stay radioactive for thousands of years and must therefore be buried in safe locations.

Despite the problems with nuclear energy, it *does* remain a source of energy that does not pollute the atmosphere as oil and coal do. As a result, the "nuclear option" may once again look attractive in the future.

ACTIVITY

Obtain a geiger counter, an instrument used for determining radioactivity. (Perhaps your school or the local university might lend you one.) Pass it over various samples of rock, including a sample that contains small amounts of a radioactive element such as uranium.

Conservation. Conservation is not often thought of as an energy resource, but it is actually the least expensive and least polluting resource we have. If more people practiced conservation, our other energy resources would last longer and there would be less pollution.

23. Referring to Figure 9.7, name five ways that you could practice conservation where you live.

24. Conservation also means making changes in your life style to conserve energy. This means driving smaller, more fuel-efficient cars or using mass transportation such as buses or trains. How would the simple life-style change of living closer to work enable you to save energy?

ACTIVITY

With a bucket and a watch, measure how many gallons of water come out of your shower in a minute. If a shower head were installed that used only two gallons per minute, how many gallons of water could you save in a seven-minute shower?

Family Tips for Saving Energy

FIGURE 9.7   Energy Saving Tips (Courtesy Boy Scouts of America)

Other Energy Sources. Many energy sources that are not widely used today may see more use in the future. Among these sources are geothermal energy, nuclear fusion, and solar and wind power.

Geothermal Energy. This energy is from the hot interior of the earth. Although the amount of energy present is enormous, it can be profitably used in only a few areas. These are areas where the heat lies close to the surface, such as near volcanos.

Nuclear Fusion. This energy source differs from nuclear fission in that it involves fusing atoms together instead of splitting them apart. When a fusion reaction occurs, tremendous amounts of energy are released. Creating this reaction is no simple process, however, and to date it has not been achieved on more than an experimental scale. If it ever does succeed on a large scale, this could be a very important energy source.

25. Nuclear fusion takes place in the cores of most stars, causing them to shine. Where in our own solar system is nuclear fusion currently taking place?

Solar Energy and Wind Energy. Solar energy is currently used in some areas for hot water and home heating. Along with wind power, it will probably see more use in the future for generating electricity as well. Both solar energy and wind energy are essentially nonpolluting. The only degrading of the environment that occurs is from mining raw materials for the manufacture of solar collectors and windmills.

26. Of the two types of recreation shown in Figure 9.8, which one is nonpolluting to the environment?

FIGURE 9.8 Modes of Recreation (Photo Researchers, Inc.)

27. What fuel still used today was widely available and burned by the early settlers for heat and light?

28. Hydroelectric power and solar energy are considered renewable resources. That is, they can be used over and over again. Oil, natural gas, and coal are considered nonrenewable resources. What does this mean?

WASTE DISPOSAL

Our society generates wastes of all sorts. These wastes are in the form of solids, liquids, and gases. If not properly disposed of, such wastes will cause pollution. Let's look at the various types of wastes we generate.

Solid Waste

Solid waste is just that—waste that is in solid form. Much of this waste is generated by homes and businesses. Figure 9.9 shows a breakdown of this waste.

In the past, solid waste was disposed of in town dumps. These dumps usually had a foul odor and served as a habitat for rodents such as rats. In addition, fires were a common nuisance at such places. As a result, many municipalities switched to landfills as a means of disposing of their solid waste (Figure 9.10).

Landfills are preferable to the old way of simply dumping trash. One problem landfills *do* have, however, is with the leachate they generate. *Leachate* is a bright orange liquid that drains from the landfill. It forms when rainwater percolates down through trash and picks up impurities, particularly iron oxide (rust). If left untreated, leachate will pollute a stream.

50% paper
12% glass
10% metal
10% food wastes
5% plastic and rubber
3% wood, dirt, grass, and leaves

FIGURE 9.9 Components of Solid Waste

173

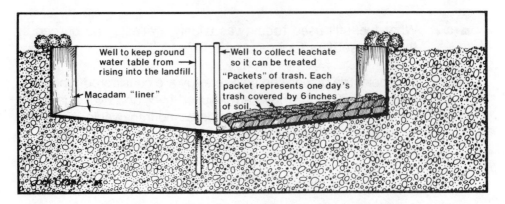

Well to keep ground water table from → rising into the landfill.

← Macadam "liner"

←Well to collect leachate so it can be treated

"Packets" of trash. Each packet represents one day's trash covered by 6 inches of soil.

FIGURE 9.10   Cross-section of an Ideal Landfill

29. When some landfills reach their capacity, a plastic cover is put over them. A layer of dirt a few inches thick is then put over this plastic cover. How do you think a plastic cover could drastically reduce the amount of leachate coming from a landfill?

30. Many towns and cities are facing the problem of their landfills filling up too rapidly. To remedy this, some municipalities have recommended recycling their solid waste. From Figure 9.9, list the waste products you think could be recycled. Then, beside each item, list its percentage of the total amount of solid waste.

Liquid Waste

The most common liquid waste our society produces is sewage. In rural areas, sewage from households is usually disposed of in septic systems that have underground drainage fields. In municipal areas, this sewage is collected and transported in sewers. These sewers converge at one central location where the sewage is treated before being released to a nearby river.

Sewage can receive three types of treatment—primary, secondary, or tertiary. Primary treatment removes the solids by means of screens and settling tanks. With this type of treatment, approximately 60 percent of the impurities are removed from sewage. With secondary treatment, approximately 90 percent of the impurities are removed. This is done by first applying primary treatment, then using bacteria to break down the remaining wastes. Secondary treatment is the method now employed by most municipalities (Figure 9.11). With

FIGURE 9.11 A Sewage Treatment Plant, Bethlehem, Pennsylvania
(Photograph by the author)

tertiary treatment, the wastes are further treated so that the outflowing water is approximately 98 percent pure. This method of treatment is expensive, however, and is not widely used.

31. Some coastal cities transport sewage wastes out to sea by barge. These wastes are then dumped. What danger might result if they are dumped too close to shore?

In addition to sewage, liquid wastes are generated in other ways also. For example, many industries generate chemical wastes unique to that particular industry. These wastes can sometimes be considered hazardous, making their disposal that much more difficult.

32. Water is used as a coolant for many industrial operations. If it is drawn from a stream, it is usually returned to the stream in much the same chemical condition. Yet this return water may readily kill fish and other organisms living directly downstream. In what way has this water changed?

175

Waste Gases

Waste gases, when not properly disposed of, cause air pollution. All of us have experienced air pollution at one time or another. Factories, automobiles, home furnaces, and power plants are some of the major sources of this type of pollution. Because of the wide variety of substances that enter the air, the problems of air pollution can be quite complicated and not easily solved. Let's look at acid rain and smog, two major problems associated with air pollution.

Acid Rain.   Acid rain forms as a result of the burning of coal and oil. These fuels contain sulfur in small amounts. When they are burned, this sulfur reacts to form a weak sulfuric acid in our atmosphere. This dilute sulfuric acid then falls to the ground as acid rain and corrodes buildings, pollutes lakes, and disrupts ecosystems.

Acid rain can be controlled by installing scrubbers in smokestacks. These scrubbers remove sulfur before it can be released to the atmosphere. They are expensive to install and operate, however, and as a result industries are sometimes reluctant to use them.

33. The world has much greater reserves of coal than of oil. This has led several people to suggest that power plants switch to burning coal instead of oil. But coal, as a rule, has more sulfur in it than oil. Pertaining to acid rain, what would be the likely result of more power plants switching to coal?

Smog.   Smog forms when air pollutants enter the atmosphere and react with each other. The sun greatly aids these reactions, making smog worse on a sunny day, particularly in the summer.

Smog can be seen over a city on a day when there is little or no breeze to blow the pollutants away. If a stagnant air mass stays over a city for several days, smog can become quite a problem.

34. Why do mass transit systems, such as trains and buses, tend to reduce air pollution in a city from what it would be if we didn't have these systems?

SUMMARY

In the House of Science, environmental scientists act as good citizens by making sure items in the house are recycled. They also work with scientists in other fields, pursuing such things as ways to heat the house more efficiently and with less pollution. In their own area of study, these scientists act as earth "managers," searching for ways to better take care of our planet.

As we look to the future of the earth, we must keep in mind the three cornerstones of the environmental triangle—population, resource use, and waste disposal. The dangers of a rapidly rising population can be seen in *The Tragedy of the Commons*, in which conditions that were good for a few grazing animals turned bad for all when the field became overpopulated.

As the second leg of the environmental triangle, natural resources include both the mineral resources and energy resources we use. Mineral resources can be classified into two groups—metals and nonmetals. Energy resources include oil and natural gas, coal, hydroelectric power, nuclear power, geothermal energy, and solar energy. To avoid running out of these resources in the future, we will have to look to alternative sources or learn to practice conservation.

As the third leg of our environmental triangle, waste disposal deals with the wastes we generate. These wastes can be classified as solids, liquids, and gases. Disposal of such wastes can pose problems to our environment.

35. Nehru, the Prime Minister of India from 1947 to 1964, once said, "Population control cannot solve any of the world's problems, but none of the world's problems can be solved without population control." What is one problem that is present in our society today, and what effect would a lower population have on that problem?

ANSWERS

1. Resource use increases as does the amount of wastes we generate.

2. It might spoil before it gets there.

3. 1900—1.6 billion
   1930—2 billion
   1960—3 billion
   1976—4 billion
   1990—5 billion

4. There has been an "explosive" growth in the number of people on the planet.

5. (a) $65.2 - 56.2 = 9$ million
   (b) $97.4 - 56.4 = 41$ million

6. $65.2 \times 1.024 = 66.76 = 66.8$ million

7. 39 years

8. **(a)** Kenya (4.7%/yr. growth rate) and Ethiopia (3.4%/yr. growth rate)
   (b) Denmark (− 0.4%/yr. growth rate)

9. There will be 1.6 percent more people next year than there are this year.

10. $90/7.1 = 12.7$

11. Eventually, there would be so many sheep that once again the grass would be eaten faster than it could grow.

12. 8.3 million

13. India—.8 billion/5 billion
    =.16 = 16 percent
    China—1.1 billion/5 billion
    =.22 = 22 percent

14. A small population increase in an industrial (developed) nation could translate into large increases in the amount of resources used and wastes generated.

15. **(a)** It is cheaper because material does not have to be transported to the surface. It is safer because there is no danger of cave-ins.
    (b) By dredging

16. Supplies are running out!

17. Plastics mainly (also fiberglass, rubber, and others)

18. Supplies cannot be replenished as farm crops are.

19. Oil floats on top of water. Pumping water into well A will cause the oil above it to float to a higher level.

20. Heating oil, gasoline, manufacture of plastics and synthetic fabrics, as a lubricant, and so on.

21. To prevent a cave-in

22. River levels may be so low that too little water is available to turn the turbines.

23. See Figure 9.7 (many answers possible).

24. Living closer to work would decrease the amount of energy you would need to get back and forth.

25. In the center of our sun

26. Sailboating

27. Wood

28. Once they are used up, they are gone and can't be replenished.

29. The plastic cover would prevent rainwater from percolating through the landfill and becoming leachate.

30. Paper—50 percent; glass—12 percent; metals—10 percent. *Note:* Some of the remaining solid waste can also be recycled, but it may not be practical to do so.

31. They may wash back to the beaches.

32. It is much warmer than it was.

33. More acid rain would be produced.

34. When trains and buses are used, fewer cars are driven. (One bus carrying 30 people pollutes less than 30 individual cars.)

35. Overcrowded schools, pollution, crime, and so on (many answers possible).

Chapter 10

GEOGRAPHY

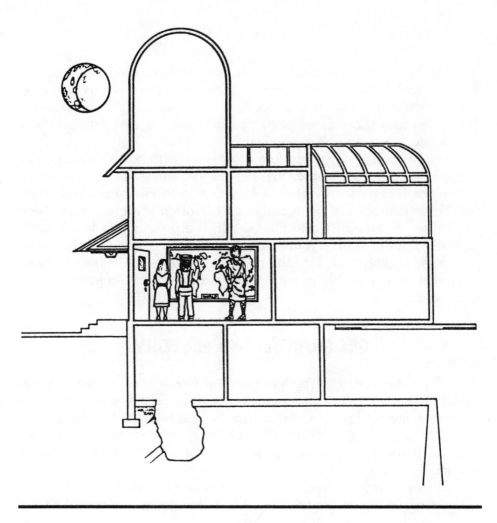

Geographers are generalists in the House of Science because their work overlaps with many other sciences. Geographers study the earth, oceans, atmosphere, and all living things on the planet. They collect and catalog information about the planet, then present this information in the form of maps, charts, or tables.

In the House of Science, the work geographers do deals with the earth and its life. These scientists examine the various lands on which we live, as well as the seas and atmosphere that surround us. They also study the various plants and animals we live with and how they relate to our life here on earth.

The work of geographers often overlaps with other sciences. For example, at any one time the geographer may act as geologist, oceanographer, meteorologist, biologist, or even environmental scientist! These many "hats" geographers wear often serve to make them generalists. A *generalist*, you may recall, is a person who knows a little about a lot of things. And just as we need specialists in science—people who know a lot about one subject—we likewise need generalists who are able to view the overall picture. Geographers are such people.

GEOGRAPHY AND HISTORY

One of the earliest geographers was the Greek poet Homer. In his book the *Odyssey* (written around 900 B.C.), he talked of foreign lands ruled by a one-eyed giant and of distant islands inhabited by mermaids. These stories, along with his tales of war in the *Illiad*, provided an exciting (if not always factual) geography lesson for students in ancient Greece.

Homer's view of world geography saw the earth as a flat disk with Greece at the center (Figure 10.1). Other geographers of the time

FIGURE 10.1 Homer's View of the World (From The Odyssey of Homer as retold by Barbara Leonie Picard. Published by Oxford University Press 1952 © Oxford University Press 1952.)

likewise saw the earth as flat, although they were likely to put their own countries at the center instead of Greece. This idea of a flat earth prevailed even until the days of Columbus with people cautioning that explorer against sailing too far from land, lest he fall off the edge of the earth!

1. When Columbus set off in search of another route to the Orient, what did he find instead?

As travel became more widespread, so did our knowledge of geography. One of the explorers who contributed to this knowledge was the Portuguese navigator Ferdinand Magellan. In 1519, he set out with five ships on an expedition to sail around the world (Figure 10.2).

Although Magellan's voyage was well-equipped at the beginning, it was ill-fated in many ways. Starvation faced the sailors during parts of the trip and Magellan himself was killed by unfriendly natives in the South Pacific. Nonetheless, on September 6, 1522, nearly three years later, one lone ship limped back to port in Portugal, thus proving to all that the earth was indeed round!

2. As shown in Figure 10.2, in which direction did Magellan travel around the world—eastward or westward?

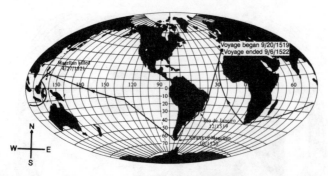

FIGURE 10.2 Magellan's Trip around the World (From *Introductory Oceanography* by Harold V. Thurman. Copyright © 1988 Merrill Publishing Co., Columbus, OH. Used with permission.)

Our knowledge of geography increased as countries such as England, France, and The Netherlands established empires around the world (Figure 10.3). Navigating around these empires required a great deal of geographical knowledge and, as a result, better maps were developed.

3. At one time it was said that, "The sun never sets on the British Empire." Referring to Figure 10.3, what was meant by this?

Today, with the aid of books, television, and films, we know quite a lot about the geography of our planet. In addition, we are well-informed about daily events in other parts of the world. But although we may know more about the world than past civilizations did, our challenges today are greater also. Among these challenges is using our knowledge to prevent destruction of the planet by means of nuclear warfare or pollution.

4. Events of worldwide interest, such as the Olympic games, can be seen in most places on earth, even as they are happening live. What modern communications device enables these television waves to be relayed to different parts of the globe?

PRESENTING INFORMATION

Geographers become adept at cataloging and presenting the information they gather. They strive to present this information in ways that can be easily understood. Maps are one way of doing this. Graphs and

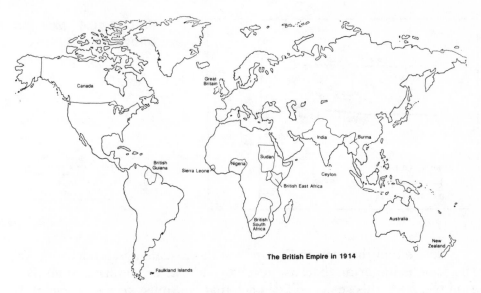

The British Empire in 1914

FIGURE 10.3   The British Empire in 1914

tables are other useful ways of presenting data. Let's look at some of these methods geographers use.

Maps

A catalog of elevations at various locations would make for quite a thick book—and not an interesting one either! Instead, geographers present this information in the form of a map. On such a map, we see the elevations as mountains and valleys and are thus able to comprehend a lot of information in a short time.

In addition to elevation, maps are made to show any number of things. A typical world map, for example, shows the way our world is divided up politically, that is, where the various countries are located. Some maps show such things as types of vegetation and soils, average rainfall, or population density. All of these maps represent a large amount of data presented in an easily readable form.

5. When you take a trip by car, what does the map you use primarily show?

To understand maps, you must first be able to read one. Figure 10.4 is a relatively simple map that shows how to get to a picnic. Directions accompany this map, although an experienced map reader would not need such directions.

183

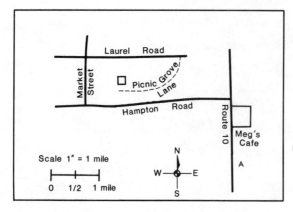

FIGURE 10.4   Map and Directions to a Picnic

Directions: Going north on route 10, make a left onto Hampton Rd. Go west on Hampton Road about 3 miles until you come to Market St. Go right on Market St. and continue north until you come to Laurel Rd. Make a right on Laurel Road and continue until you come to Picnic Grove Lane.

Notice that the map in Figure 10.4 has a scale and a north arrow. The scale is important because it tells you how far you have to go. For example, from this scale, you can see that you must go about one mile on Market Street before you reach Laurel Road. And from the north arrow, you can tell that you must travel north.

6. As you travel northward on Route 10, what landmark would you look for to tell you that you were getting close to Hampton Road?

7. About how far must you travel on Laurel Road before you come to Picnic Grove Lane?

8. Traveling east on Laurel Road, do you make a right or a left onto Picnic Grove Lane?

9. The expression "as the crow flies" refers to the flight path of a crow (or any bird). It is generally taken to mean a straight path from one point to another. What is the approximate distance between point A (just south of Meg's Cafe) and the Picnic Grove "as the crow flies"?

10. If you were to walk in a straight line from point A to the Picnic Grove, in what direction would you be walking?

The world map shown in Figure 10.5 is in some ways like the map shown in Figure 10.4. It has a north arrow and a scale, although the scale is much larger because this map shows the entire earth instead of only a small portion of it.

184

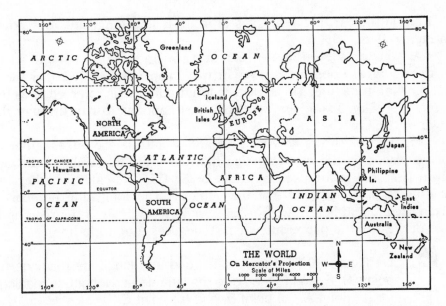

FIGURE 10.5   A World Map—Mercator Projection (Copyright © by D.C. Heath and Company.)

You would do well to buy a large map of the world similar to that shown in Figure 10.5. Every day memorize a country, island, or body of water. In this way, you will gradually build up an understanding of world geography. This understanding will help you in many other subjects as well.

11. Antarctica is the only continent not shown on Figure 10.5. That is because it lies over the South Pole and does not project well on this type of map. What six other continents *are* shown on this map?

12. In addition to the Antarctic Ocean, four other major oceans cover our planet. Name them.

13. What continent is located directly south of Europe?

14. What continent (other than Antarctica) is located entirely south of the equator?

15. What ocean separates North America from Europe and Africa?

Latitude and Longitude.   To locate a place on the earth, a system of coordinates is needed. The system we use is based on imaginary lines that circle the earth. These lines of latitude and longitude are shown in Figure 10.6.

As you can see in Figure 10.6, latitude lines run north and south of the equator. Longitude lines, in turn, run east and west of the 0 degree longitude line. This 0 degree longitude line is referred to as the *prime meridian*. The prime meridian just happens to run through the town of Greenwich, England.

Figure 10.7 shows how latitude and longitude are put together to locate places on the earth. Notice that the latitude number is always given first and is followed by an N or S. This indicates whether the reading is north or south of the equator. The longitude number is given second and is followed by an E or W to indicate if the reading is east or west of the prime meridian (0 degrees longitude line).

16. On Figure 10.7, study the latitude and longitude readings for points D, E, and F, then give the latitude and longitude readings for **(a)** point A, **(b)** point B, and **(c)** point C.

17. The *international date line* corresponds to a line of longitude located halfway around the world from the prime meridian. If there are 360 degrees in a circle, how many degrees away from the prime meridian is the international date line?

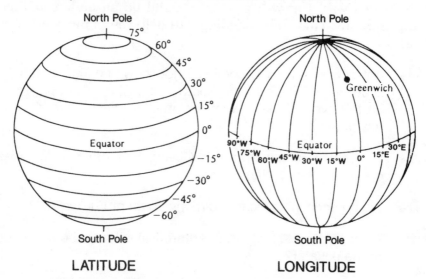

FIGURE 10.6   Latitude and Longitude Lines (From *Horizons: Exploring the Universe*, 1987 Edition, by Michael A. Seeds © 1987 by Wadsworth, Inc. Reprinted by permission of the publisher.)

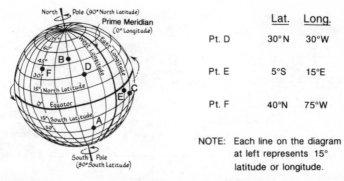

	Lat.	Long.
Pt. D	30°N	30°W
Pt. E	5°S	15°E
Pt. F	40°N	75°W

NOTE: Each line on the diagram at left represents 15° latitude or longitude.

FIGURE 10.7   Locating Places on the Earth by Latitude and Longitude (From *View of the Earth, An Introduction to Geology* by John J. Fagan. Copyright © 1965 by Holt, Rinehart and Winston, Inc. Reprinted by permission of the publisher.)

18. When a new day begins just after midnight at the international date line, what time is it at the prime meridian?

ACTIVITY

Working with a partner, give latitude and longitude readings for a particular landmark on the globe. See if your partner is able to find the place you are thinking of. Take turns doing this.

Map Projections. A globe is the most accurate way of presenting the geography of the world. However, it is not always convenient to transport a globe from place to place. We therefore rely on the convenience of flat-lying maps.

To make a map of the world, we might imagine "skinning" the outside of a globe much as we would skin the peel off an orange. This "skin" of the world is then stretched out and laid flat so that what we see is a projection of what our earth looks like. But just as you would have to stretch an orange peel in places to get it to lie perfectly flat, you must likewise stretch the skin of the globe in places to make it lie flat. And this stretching leads to the distortion of areas from what they look like on the globe. Figure 10.8 shows a projection of the earth in which distortion has been minimized. You might imagine that if you cut this map out and fit it together, it would make a pretty good globe.

19. If you were studying ocean areas of the world, why would the map projection shown in Figure 10.8 not be a good one to work with?

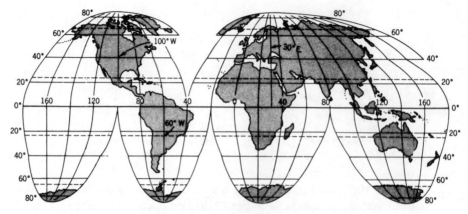

FIGURE 10.8   One Projection of the World

A familiar projection used by geographers is the *Mercator projection*. In this projection, the skin of the globe is cut from pole to pole along one of the lines of longitude. Like a piece of rubber, it is then pulled out flat so that it forms a perfect rectangle. To perform this feat, areas near the poles must be stretched a great deal, while areas near the equator are stretched very little. As a result, areas near the poles appear much larger than what they actually are. Figure 10.5 is a map of the world using the Mercator projection.

20. (a) Greenland, located northeast of Canada, is a land mass that in reality is only about one-eighth the size of South

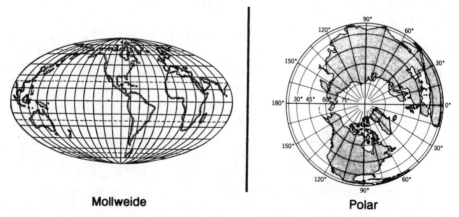

Mollweide   Polar

FIGURE 10.9   Mollweide and Polor Map Projections (Left: Copyright © by D.C. Heath and Company. Right: From *Introduction to Physical Geography* by Arthur N. Strahler. Copyright © 1965, 1970, by John Wiley & Sons, Inc.)

America. On the map shown in Figure 10.5, which of these two land masses appears larger?

(b) Compare Greenland to Australia on Figure 10.5. Which of these two land masses do you think is larger in reality?

21. In Figure 10.5, why might Alaska, northern Canada, and much of the Soviet Union appear larger than they actually are?

In an attempt to better portray areas of the world, other map projections have been developed. These projections also distort some areas, but not as much as the Mercator projection does (Figure 10.9).

22. Where do all the longitude lines meet on the Mollweide projection?

23. Which southern continent do you think is usually studied with maps having a polar projection?

24. **(a)** What type of projection is shown in Figure 10.2?

(b) What type of projection is shown in Figure 10.3?

Other Ways of Presenting Information

In addition to maps, geographers have other ways of presenting information. They sometimes present it in the form of a table, for example. In the previous chapter, Tables 9.1 and 9.2 are examples of this method of presenting information. In these tables, population data are being presented.

Graphs are another way of cataloging and presenting data so that it can be easily understood. Figure 10.10 is such a graph. It shows oil production for the various countries of the world.

25. **(a)** After the Organization of Petroleum Exporting Countries (OPEC), which country produces the most crude oil?

(b) Which country is the next highest producer?

Figure 10.11 shows another way of presenting geographical data. This picture graph method can be very useful because it lets you actually visualize the information being presented.

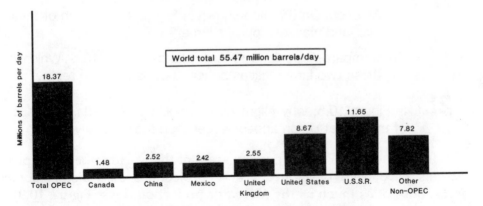

FIGURE 10.10   World Production of Crude Oil, 1986 (*Source:*
Energy Information, Annual Energy Review, 1986.)

26. The vegetation changes shown in Figure 10.11 are due mainly
to the cooler temperatures that exist at higher elevations.
These cooler temperatures prevent certain types of vegetation
from getting established. At about what height do the first evergreen trees appear?

27. At about what height does the icecap begin?

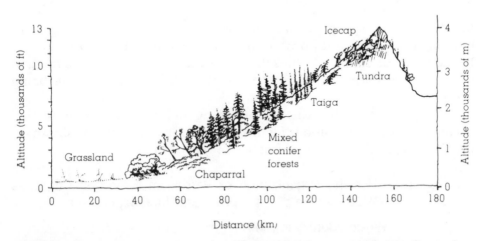

FIGURE 10.11   **Graph of Changing Vegetation Due to Altitude—Central
Sierra Nevada Mountains** (Reprinted by permission from *Meteorology Today*
by C. Donald Ahrens. Copyright © 1982 by West Publishing Company. All
rights reserved.)

GEOGRAPHY AND PEOPLE

Geography is more than just collecting and presenting information about the planet, although this part of it is important. Geography also looks at how people live in their environment, be it a tropical rain forest, a desert, or simply a humid climate such as many of us experience.

How people make a living in their environment is a part of what we refer to as their *culture*. Culture also takes into account such things as the language people speak, their manner of dress, and the types of food they eat. Culture, in turn, is strongly affected by the environment people live in.

ACTIVITY

Obtain an inexpensive sampler of stamps (or coins) from other countries. By examining them, try to determine what is important in the cultures of these countries—famous people, flowers, athletes?

Figure 10.12 shows various types of shelter that were at one time built by native peoples. Today, houses from different parts of the world often look much the same. But, in the past, it is interesting to see how people used their environment to construct the homes they lived in.

28. **(a)** Which type of shelter shown in Figure 10.12 would likely be used by nomads in the desert?

(b) Why is this type of shelter practical for their way of life?

29. Referring to Figure 10.12, which type of shelter would be used where protection is needed against crawling insects?

30. The early settlers in North America built log cabins because the wood was readily available. What did Eskimos use to build their houses and why did they use this material?

APPLIED GEOGRAPHY

Geographers apply their knowledge in a wide range of areas. They apply it in laying out the route of a new highway, for example, or in locating a store in an ideal spot for business. On another scale, they apply it in a foreign country with an organization such as the Peace Corps.

One practical application of geography is in the field of land-use planning. As the name implies, this type of work involves planning the future use of land. To do this, planners must work with information that has been collected, particularly maps. Good planners also work

A

B

C

D

E

F

FIGURE 10.12   Types of Shelter

with the people in an area, getting to know something about their culture and way of life. Let's look at an actual example of the type of work that planners do.

Figure 10.13 shows the town of Hockessin which is located in northern Delaware. This area has rich farmland and is characterized by rolling hills. It has a temperate climate, and the average rainfall is about forty inches per year. Many of the people living there work in the nearby city of Wilmington, Delaware.

31. The road map shown in Figure 10.13 is indispensable to a planner for locating features in the area. This road map also serves as a base for other maps of the area. Two major roads run through the Hockessin area. They each have a route number. What are the names of these roads and their route numbers?

32. Sewer lines have been laid in the Haverford and Gateway Farms developments (Figure 10.13, lower part). Sewers are needed whenever there are a lot of houses in an area. Other housing developments in this area will need sewers in the future. Name two of these housing developments.

Large developments generally switch over to a central water supply, rather than providing each home with its own well. Before planning a water system for a development, it is first necessary to

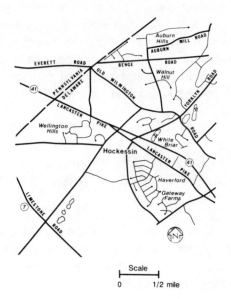

Scale

0   1/2 mile

FIGURE 10.13   Road Map of the Hockessin Area

193

determine how much water that development is going to need. To determine this we can use this formula:

Water Use = Number of Houses in Development ×
4 people/house × 100 gallons of water/day

Notice that we use the average figures of four people per household and 100 gallons of water per person per day. These are good average figures.

33. How many gallons of water per day will the Haverford development require if there are 275 houses located in that development?

Figure 10.14 is a geologic map of the area. This map shows the type of rock found at the surface, or if soil covers the area, the first rock type that would be found if that soil cover were stripped away. As you can see, there are two major rock formations underlying the Hockessin area—the Cockeysville Marble and the Wissahickon Formation.

34. If you were to drill one well to provide for the water needs of the Haverford development, which formation would you choose to drill into and why?

35. Water use fluctuates in the course of a day. There are times of peak demand, such as at breakfast and supper, and times of little demand, such as in the middle of the night. A well might be able to handle a housing development's overall demand for water, yet be unable to handle its peak demand. Instead of drilling a second well, however, a storage tank could be built. How might a storage tank enable a well to handle the peak demand for water?

The soils map shown in Figure 10.15 is useful to many individuals, particularly farmers and builders. Builders use soils maps to determine whether an area is likely to be suitable for septic systems. They also use them to determine if a house is likely to have a wet basement or not.

36. In Figure 10.15, those soils that begin with the letters C or G are generally suitable for septic systems, whereas those beginning with the letter H are not. List those soils that are suitable for septic systems and those which are unsuitable.

 Wissahickon Formation (wells drilled into this formation average 10 gallons per minute)

 Cockeysville Marble (wells drilled into this formation average 100 gallons per minute)

FIGURE 10.14 Geologic Map (Base map: U.S. Geological Survey. Geology overlay: Delaware Geological Survey)

37. Notice that Figure 10.15 is also an aerial photograph of the area. From this photograph, what would you estimate occupies more area—woods or farmland?

In looking to the future of the Hockessin area, planners must consider how future development will affect things such as the runoff of rainwater following storms. A large shopping center, for example, will cause water to run off the land instead of soaking into the ground. As a

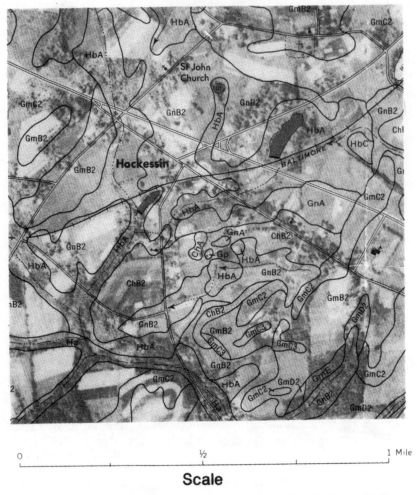

Scale

FIGURE 10.15 Soils Map (U.S. Department of Agriculture, Soil Conservation Service)

result, this type of development could cause the flooding of some areas following heavy storms.

38. In many urban areas of the country, floods of today are worse than floods of the past, even though rainfall has not changed appreciably. Why might this be?

39. Residents in a locality are sometimes divided as to what their community's future should be. If a shopping center is to be built, for example, some of the residents may be for it, while

196

others are against it. Give one reason why residents might like to see a shopping center built in their community. Give one reason why they might not.

SUMMARY

In the House of Science, the work geographers do often overlaps with other sciences. As such, geographers may be thought of as generalists; that is, they know something about a lot of the sciences.

Our knowledge of geography has gradually accumulated throughout history, with people such as Homer, Columbus, and Magellan adding to our store of information. Today, much of geography consists of collecting and cataloging information about the planet. This information is then presented in a form that can be easily read, such as on maps, graphs, or tables. But geography is more than just collecting and presenting data. It also looks at how people use their surroundings to make a living. It examines what we refer to as their culture.

Geographers apply their knowledge in many fields of endeavor, such as in the Peace Corps and in land use planning.

40. When thinking about the world, we tend to think of everyone going through their day just as we are. But as people in North and South America are "up and about," most of the people in Asia and Australia are sleeping. Why is this?

ANSWERS

1. The New World (North America)

2. Westward

3. The British Empire encompassed so many lands that it was always daytime in at least one of them!

4. The satellite

5. The roads

6. Meg's Cafe

7. About two miles

8. A right

9. About three miles

10. Northwest

11. Africa, Asia, Australia, Europe, North America, and South America

12. Atlantic, Pacific, Indian, and Arctic

13. Africa

14. Australia

15. The Atlantic Ocean

16. Point A: lat. 30°S, long. 30°W
    Point B: lat. 50°N, long. 45°W
    Point C: lat. 15°S, long. 40°E

17. 180 degrees

18. Just after noon

19. The map is not continuous over
    the ocean areas. Instead, it is
    "sliced" over these areas.

20. (a) Greenland
    (b) Australia is in reality more
        than twice the size of
        Greenland.

21. These areas all lie fairly close to
    the North Pole, and areas near the
    poles are greatly exaggerated on
    this projection.

22. At the poles

23. Antarctica

24. (a) Mollweide
    (b) Mercator

25. (a) U.S.S.R.
    (b) United States

26. About 3000 feet (1000 m)

27. About 10,000 feet (3200 m)

28. (a) Tents (d)
    (b) They can easily dismantle it
        when they move to another
        area.

29. Tree house (c)

30. They used ice because it was
    readily available.

31. Route 7, Limestone Road; Route
    41, Lancaster Pike

32. Wellington Hills, White Briar,
    Walnut Hill, Auburn Hills—any two

33. 275 houses × 4 people/house × 100
    gallons per person/day = 110,000
    gallons/day

34. The well should be drilled into
    the Cockeysville Marble because
    this formation is more likely to
    yield large quantities of water.

35. Water could be pumped into it
    during nonpeak times and released
    during peak times when the well
    cannot keep up with demand.

36. Suitable: ChB2, GmB2, GmC2,
    GmD2, GmE, GnB2
    Unsuitable: Ha, HbA, HbC

37. Farmland

38. More areas are paved today than
    were in the past.

39. For: convenient shopping, tax
    revenues
    Against: increased traffic, noise,
    unsightly, higher chance of
    flooding

40. They are located halfway around
    the world and it is nighttime there.

Chapter 11

ASTRONOMY

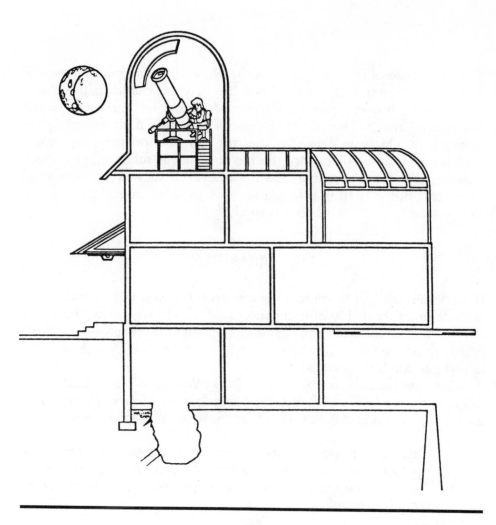

Astronomers work in the observatory of the House of Science. Using telescopes and the spectroscope, they study distant planets, stars, and galaxies in an effort to better understand our universe. Today, we know more about space than was ever imagined.

I n the House of Science, most of the rooms are occupied by people working to better understand our life here on earth. In the observatory of the house, however, astronomers look past our earth to the frontiers of space. They do so in hopes of better understanding the universe in which we live. To date, the information they have gathered has given us a picture of the heavens that is much more encompassing (and mind-boggling) than was ever imagined by people of past ages. Some of this information has even changed the way we view our very existence here on earth.

THE UNIVERSE

If you could take an "inventory" of our universe, you would find that it is made up of about 100 billion galaxies, of which our galaxy, the Milky Way galaxy, is but one. These galaxies, in turn, are each made up of *billions* of stars. Our Milky Way galaxy, for example, is thought to contain about 200 billion stars.

Because we are a member of the Milky Way galaxy, it is hard for us to get an "outsider's view" of what our galaxy looks like. Figure 11.1, however, shows a galaxy that is in many ways similar to our own.

1. Just as the earth revolves around the sun, the stars in a galaxy revolve around its center. As seen from earth, which way are the stars in the M74 galaxy revolving, clockwise or counterclockwise?

200

FIGURE 11.1 Galaxy M74, a Galaxy Similar to Ours (Note: The white arrow represents where our sun would be located if this were our own Milky Way Galaxy) (Lick Observatory Photograph)

2. Our sun is just one star in the Milky Way galaxy. About how many other "suns" are in our galaxy?

ACTIVITY

Visit a planetarium, an indoor universe. Note some of the various constellations, such as the Big Dipper and Orion. Note also how the constellations in the northern hemisphere seem to revolve around our North Star.

On earth we measure large distances in terms of miles or kilometers. In astronomy, we use a different unit of measure because the distances involved are so much greater. One unit we use is called the *light year*. As you might guess, a light year is the distance that light travels in one year. Remembering that light travels at 186,000 miles *per second* (300,000 kilometers/sec.), you might also guess that the distance light travels in one year would be truly large—and it is—5.8 *trillion* miles (9.3 *trillion* km)!

Although light travels quite rapidly, the distances involved in astronomy are quite large also. Looking at the size of galaxies, for example, we find that our own Milky Way galaxy has a diameter of approximately 100,000 light years. This means that if you were to shine a light from a space station at the very edge of our galaxy, it would take 100,000 years for that light to reach the other edge of the galaxy!

3. What is the diameter of our galaxy in trillions of miles (km)? (*Hint*: Multiply light years by miles (km) per light year.)

4. One of the stars closest to our sun is called Barnard's star. It is thought that this star may have planets circling around it, just as our sun has planets that circle it. Barnard's star is four light years away from us. If we were to send a spacecraft toward Barnard's star at one-quarter the speed of light (not something we will likely accomplish soon), how long would the entire round trip take, assuming we found life there and stayed for a year?

Approximately 20 galaxies exist within three *million* light years of our own Milky Way galaxy. This "family" of galaxies, of which we are a member, form a cluster known as the *Local Group*. In the Local Group, the Andromeda galaxy is our nearest galactic neighbor, being two million light years from us (Figure 11.2).

5. It is said that we are seeing the Andromeda galaxy as it was two million years ago, not as it is today. Why is this?

6. If intelligent beings from the Andromeda galaxy were observing our galaxy, and in particular our earth, they might be able to observe life on the planet. Yet they would see nothing resembling modern human life. Why not?

FIGURE 11.2 The Andromeda Galaxy (Note: The two bright spots with arrows indicate dwarf galaxies near the Andromeda galaxy) (Lick Observatory Photograph)

Rounding out our "tour" of the universe, we find that the closest galaxies to our Local Group are approximately 30 million light years away. At the other extreme, the farthest galaxies observed are thought to be approximately two *billion* light years away!

ACTIVITY

Divide a group of your friends, classmates, or family into two teams and hold a debate on whether or not you think there is intelligent life elsewhere in the universe.

EVOLUTION OF STARS

Just as a campfire burns fuel, a star must burn fuel to give off energy. However, the fuel that stars burn is unlike that burned in a campfire. The fuel source for stars is a process known as *nuclear fusion*. As shown in Figure 11.3, nuclear fusion consists of four hydrogen atoms fusing together to form one helium atom. When this reaction occurs, great amounts of energy are given off. It is this nuclear fusion that takes place in the cores of most stars, our sun included.

Our sun formed as other stars form, by condensing from a large cloud of dust and gas. As this cloud condensed, it heated up, and at a temperature of 10 million degrees Centigrade, nuclear fusion commenced in the core. Shortly thereafter, our sun began to shine!

It is estimated that our sun has been shining for five billion years and that it will continue to do so for another five billion. This means that it is about halfway through its life cycle (Figure 11.4).

Toward the end of our sun's life, great amounts of helium will have accumulated in its core. This core of helium gas will not be able to burn, and as a result will start to collapse. When this happens, heat will be generated, and the remaining hydrogen fuel will start to burn furiously. The sun will expand tremendously, becoming what is known

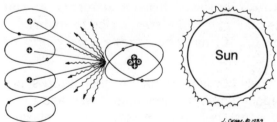

J. Crane © 1989 **FIGURE 11.3   Nuclear Fusion**

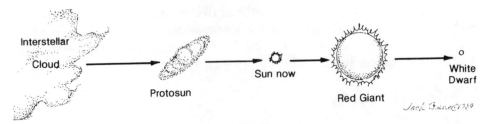

Interstellar Cloud → Protosun → Sun now → Red Giant → White Dwarf

Jack Baumer 1989

FIGURE 11.4   Evolution of Small and Medium-Sized Stars (such as our sun)

as a *red giant* (Figure 11.4). When this happens, the planets closest to the sun, including our earth, will be scorched!

After our sun has burned most of its fuel, its outer shell of gases will escape into space and what remains will be a *white dwarf*. As a white dwarf, our sun will be much smaller than it was, but will still be "white hot." As it gradually cools, however, it will eventually fade out and die.

7. How long will our sun take to go through its life cycle?

Many of the stars in the sky go through a life cycle very similar to that of our sun (Figure 11.4). Stars a good deal smaller, however, never become red giants. They simply become white dwarfs when their supply of hydrogen is exhausted. These smaller stars burn their fuel slower and thus take longer to go through the life cycle. In fact, some of these stars may take *trillions* of years to exhaust their nuclear fuel and become white dwarfs!

Like stars much smaller than our sun, those much larger than our sun go through a different life cycle. These larger stars—which account for about 10 percent of all those found in the universe—*do* share a similar beginning with normal stars, however. That is, in their cores, hydrogen atoms fuse to become helium, so that with time a helium core builds up in the star. But in these large stars, enough material is present so that the helium itself eventually begins to burn and fuse into carbon. Carbon, in turn, can fuse into even heavier elements. As you might guess, as these changes take place, the star varies in its brightness. And, as shown in Figure 11.5, it finally becomes a *supergiant*.

8. As the giant star furiously burns its remaining fuel, it turns into a supergiant, a star many times brighter than it was and thousands of times brighter than our sun! For very distant galaxies,

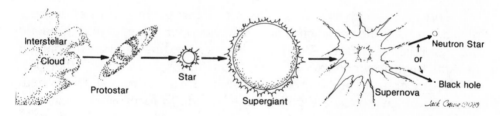

FIGURE 11.5   Life Cycle of a Large Star

why might our telescopes be able to pick out only those stars that are currently supergiants?

Following the supergiant stage (Figure 11.5), the mammoth star begins to collapse. As this collapse occurs, the pressures in its core reach staggering proportions—perhaps a trillion trillion *tons* per square inch! These pressures eventually push atomic nuclei so close together that they actually touch each other. When this happens, the star can be compressed no further and the collapse comes to a halt. Then, like a giant spring compressed to its limit, the star instantly rebounds and we see a terrific explosion called a *supernova* (Figure 11.5). During a supernova, the star appears *billions* of times brighter than it was!

9. A supernova was recorded by the Chinese in the year 1054 A.D. This exploding star became so brilliant that it was visible for several weeks, even in broad daylight! Although we receive the light from stars in the daytime, we are not usually able to see it. Why not?

When a star explodes into a supernova, the elements that were formed in its core are thrown out to the surrounding sky. During this time, some atomic nuclei are broken apart, releasing protons and neutrons. These protons and neutrons are then captured by other atoms, and in this way the elements heavier than iron (see Figure 3.3) are formed. Silver, gold, and uranium, for example, are all formed during a supernova. It is believed that this is the only way these elements form.

10. Our sun and solar system are thought to have formed from a large cloud of gas and dust particles. Much of this gas and dust is thought to have been the remains of past supernovae explosions. How does the presence of heavy elements on our planet (and elsewhere in the solar system) tend to confirm this?

Following a supernova, the large star has burned out its fuel supply and quickly fades in brilliance. What happens then is anticlimactic but nevertheless bizarre. The material surrounding the central part of the star begins to contract and this contraction is so strong that protons and electrons are forced together to form neutrons! These neutrons are in turn packed together side by side to form what is called a *neutron star*. A neutron star (also known as a *pulsar*) is only about 20 miles in diameter!

11. You may recall that atoms are composed of electrons whirling around a central nucleus. You may also recall that this nucleus is very small in comparison to the total diameter of the atom. In fact, if an atom could be enlarged to the size of a house, the nucleus of that atom would be no larger than the size of a pinhead! This holds true even for the heavier elements such as lead and gold. What is meant by the saying that an atom is largely empty space?

In a neutron star, there *is* no empty space between the nucleus and surrounding electrons. Such a star contains only neutrons fused together, side by side. At such a density, it is estimated that a teaspoon of a neutron star would weigh five *billion* tons on the earth! If you were to put a small sample of such a star on your desk, it would go through your desk, through the floor, and straight into the center of the earth!

Although many large stars eventually wind up as neutron stars, not all of them do. Some of them contain so much material that they simply continue to collapse. They collapse to form what is known as a *black hole*. A black hole is so dense that not even light can escape from its pull of gravity. That is, an object shining from its surface can never be seen by others, because the gravity of the black hole simply pulls the light rays back in!

12. Why might astronomers have trouble proving that a black hole exists in a certain part of the sky?

TOOLS OF THE ASTRONOMER

In the early days of astronomy, stargazers relied on their eyes to observe the stars. People with good vision could discern about 5,000 stars in the evening sky. They could plot the paths of planets such as Venus, Mars, and Jupiter, and they could observe the phases of the moon. They might also have observed that some areas of the moon appeared brighter than others.

In 1609, Galileo used the newly invented telescope to observe the sky. With this tool, he was able to see 50,000 stars instead of 5,000. He was also able to observe such things as craters on the moon and satellites (moons) circling the planet Jupiter.

Figure 11.6 shows a modern-day reflector telescope. Such telescopes enjoy wide use in astronomy today. Instead of lenses, such as Galileo's telescope used, they use mirrors to gather and magnify starlight. These mirrors can range in size from a few centimeters to several meters in diameter.

13. In Figure 11.6, the starlight is coming in directly over the observer's head. It is then being reflected off a large curved mirror to where he can see it. What type of reflector telescope is this, Prime focus or Newtonian focus?

ACTIVITY

Visit an observatory and look at stars and planets with a telescope. Can you see moons around Jupiter? Does Saturn have a

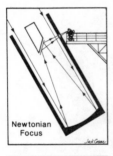

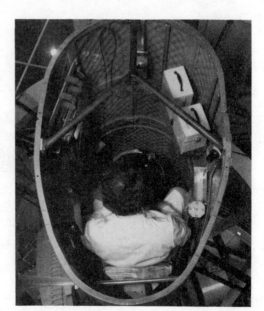

FIGURE 11.6   A Reflector Telescope (Right: Lick Observatory Photograph)

ring around it? Do some stars have a reddish tint while others shine with a bluish light? What constellations can you discern?

Some telescopes collect radiation that is not visible to our eyes. Some of this radiation is in the radio portion of the electromagnetic spectrum (see Figure 4.11). Because these instruments receive radio waves, they are called *radio telescopes*. Radio telescopes are unlike ordinary telescopes in that they contain no lenses or mirrors. Instead, they are made up of a wire mesh that reflects radio waves toward a central antennae (Figure 11.7).

14. With a radio telescope we must rely on an instrument to record the presence and intensity of radio waves. Why can't we simply observe these radio waves like we do with a reflector telescope?

Another important tool used by astronomers is the *spectroscope*. This instrument is responsible for much of what we know today about stars and how they evolve. It works on the principle of the *prism*; that is, it breaks light apart into its different wavelengths (Figure 11.8).

Just as your eye can tell the difference between the light from an incandescent lamp and a fluorescent one, a spectroscope can

FIGURE 11.7 A Radio Telescope (Note: This instrument collapsed in Autumn, 1988) (Photograph by J.R. Holzinger)

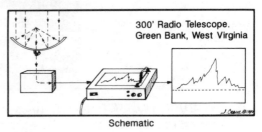

300' Radio Telescope. Green Bank, West Virginia

Schematic

FIGURE 11.8   A Spectroscope

distinguish between starlight of many types. In fact, it can tell what elements are present in a star simply by breaking apart the light it receives from that star!

15. **(a)** In Figure 11.8, what instrument is used to collect the light for the spectroscope?

(b) What instrument actually breaks the light apart?

Referring to Figure 11.8, some of the light passing through the outer atmosphere of a star is absorbed by that atmosphere. As a result, we do not see certain wavelengths. The wavelengths we don't see appear as dark lines when the light is broken apart here on earth by the spectroscope.

For any given element, the pattern of dark lines we see corresponds exactly to the wavelengths of light that were absorbed by the star's atmosphere. These particular wavelengths were absorbed due to the elements present in that atmosphere. Each element, you may remember, has a different number and arrangement of electrons in shells around its nucleus (see Figure 3.2). Because of this, each element absorbs a different set of light wavelengths, thereby producing its own unique pattern of dark lines (Figure 11.9).

As shown in Figure 11.9, the pattern of dark lines produced by any one element serves as a "fingerprint" for that element. From such "fingerprints" we can tell what elements are present in a star's atmosphere. We can then infer that these elements are probably present in the star as well.

16. In Figure 11.9, most of the spectral lines (dark lines) for the element argon occur in what color portion of the spectrum?

OUR EXPANDING UNIVERSE

When astronomers examine galaxies with the spectroscope, they find that these galaxies contain many of the same elements found in our sun and on earth. What is curious, however, is that the spectral lines are all displaced to longer wavelengths (Figure 11.10). This displacement is known as the *red shift* because the lines are all shifted toward the red part of the spectrum.

As seen in Figure 11.10, waves of light undergo a red shift if they are from an object that is moving away from us. The amount of this red shift depends on how fast the object is moving away.

17. If an object were moving toward us instead of away, the waves of light would be shifted toward the blue end of the spectrum. We would therefore see the object as giving off shorter wavelength light than it actually does. Thus, for any given element

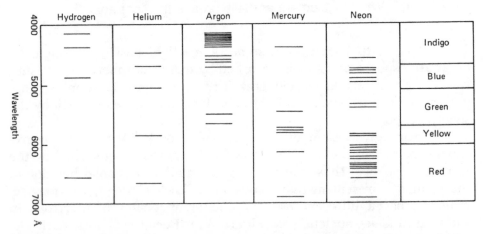

FIGURE 11.9 Spectral Line Patterns for Several Elements (Reprinted with permission of Macmillan Publishing Company from *Laboratory Exercises in Astronomy* by Joseph R. Holzinger and Michael A. Seeds. Copyright © 1976 by Macmillan Publishing Company.)

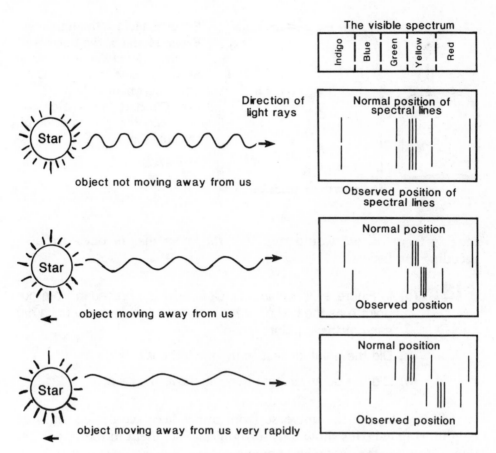

FIGURE 11.10   The Shifting of Spectral Lines

we would see a shift in the pattern of spectral lines. On Figure 11.10, would this shift be to the left or to the right of where these lines would normally appear?

Astronomers have noticed that all of the galaxies in the universe exhibit a red shift in their spectral lines. This shift is small for some galaxies and great for others. What this means is that all of the galaxies in the universe are moving away from us (and from each other as well)! To explain how this could happen, astronomers have resorted to an unlikely analogy—a loaf of raisin bread (Figure 11.11).

In Figure 11.11, the letter A represents our galaxy. This point was randomly picked and could have been represented by any of the points (raisins) in the loaf of bread. Notice, however, that as the bread rises, all of the points move apart. If we were to measure how fast they are

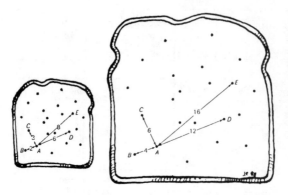

FIGURE 11.11 The Raisin Bread Model of the Universe (From *Realm of the Universe, Second Edition* by George O. Abell. Copyright © 1980 by Holt, Rinehart and Winston, Inc. Reprinted by permission of the publisher.)

moving from us, we would find that the most distant ones are also receding the fastest.

18. **(a)** In Figure 11.11, points B, C, D, and E all receded from our point A as the loaf of bread was baked. How many units did each of these points move?

(b) Did the point closest to us move the least?

(c) Did the point farthest from us move the most?

As we observe the universe from our galaxy, we notice that our neighboring galaxies show only a slight displacement in their spectral lines. More distant galaxies, however, show a greater spectral shift. In fact, the farther the galaxy is from us, the greater the shift we see. This means that the more distant galaxies are also moving away from us the fastest, as was the case with the raisins in our rising loaf of bread.

Based on the red shift we see in spectral lines, astronomers believe that our universe is expanding. Such an expansion, they believe, was caused by an event referred to as the *Big Bang* (Figure 11.12). The Big Bang is thought to have occurred around 15 to 20 billion years ago.

As seen in Figure 11.12, the Big Bang threw all of the matter in the universe violently apart. With time, much of this matter coalesced into galaxies. Today, our universe continues to expand as these galaxies fly apart in the aftermath of the Big Bang.

19. Some of the galaxies most distant from ours are moving from us at a rate of 90 percent of the speed of light! If the speed of

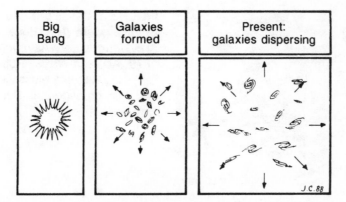

FIGURE 11.12   The Big Bang (Adapted from *Astronomy: Fundamentals and Frontiers* by Robert Jastrow and Malcolm H. Thompson. Copyright © 1972 by John Wiley & Sons, Inc.)

light is 186,000 miles per second (300,000 km/sec.), how fast are these galaxies moving away from us?

Astronomers are uncertain if the universe will continue to expand forever. If enough gravity is present, the universe eventually *will* stop expanding. In fact, it could begin to contract, so that we would see a shift in the spectral lines in the opposite direction!

20. The discovery of more black holes in the universe would strengthen the theory that one day the universe may stop its expansion. Why would black holes help stop this expansion?

OUR SOLAR SYSTEM

Our solar system came into existence about five billion years ago. It is believed to have formed from a large cloud of gas and dust particles that began to contract. As shown in Figure 11.13, this contraction caused the cloud to spin, and in so doing, flatten into a disk. Within this disk, the sun and planets began to form.

21. What term is used to describe the early planets, before they were completely formed?

Our solar system is the part of the heavens that we are most familiar with. It is our own back yard, so to speak. Figure 11.14 shows the

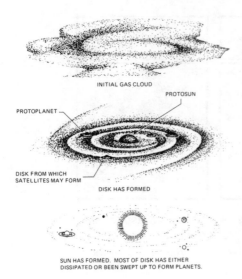

INITIAL GAS CLOUD

PROTOSUN

PROTOPLANET

DISK FROM WHICH
SATELLITES MAY FORM

DISK HAS FORMED

SUN HAS FORMED. MOST OF DISK HAS EITHER
DISSIPATED OR BEEN SWEPT UP TO FORM PLANETS.

FIGURE 11.13 Formation of the Solar System (N.A. Pananides/T. Arny, *Introductory Astronomy,* © 1979, Addison-Wesley Publishing Co., Inc., Reading Massachusetts. Fig. 9.26 on page 206. Reprinted with permission.)

planets of our solar system along with their relative distances from the sun.

22. Of the nine planets in our solar system, four are many times larger than the rest. These four are referred to as the *giant planets.* What are their names?

23. There is an area in our solar system consisting of thousands of boulders that are orbiting the sun. These boulders either never formed into a planet or are the remains of a planet that broke apart. What is the name of this zone of boulders?

ACTIVITY

Make a list of items you would include inside a spacecraft for you to survive a trip to Mars.

Our Earth

Our earth is located between Venus and Mars in the solar system. Our planet is unique in several ways. First, its distance from the sun is optimal for receiving the sun's rays. Unlike other planets that receive too much or too little sunlight, we receive just the right amount for life to exist on the planet.

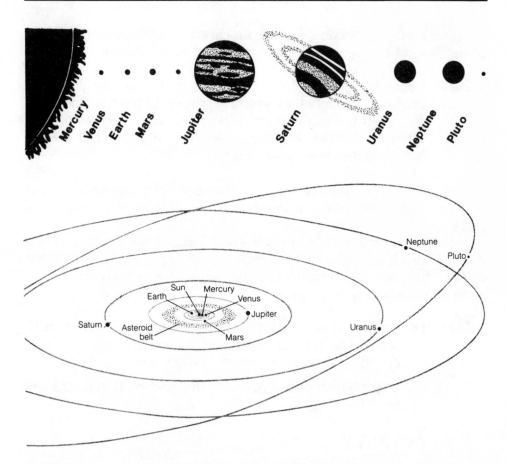

FIGURE 11.14 Our Solar System (From *Earth*, 4/E by Frank Press and Raymond Siever. Copyright © 1974, 1978, 1982, and 1986 W.H. Freeman and Company. Reprinted with permission.)

Our earth is also unique in that it has developed an atmosphere— and that it possesses enough gravity to hold onto this atmosphere. Mars, for example, is believed to have once had a much denser atmosphere than it does today. But Mars is only about half the size of earth, and its gravity was not sufficient to hold onto this atmosphere. As a result, much of it drifted off into space.

24. The planet Venus can be considered a sister planet of our earth because of its similar size and distance from the sun. There is no life on Venus, however, for its surface temperature is thought to be close to 900 degrees Centigrade! Why wouldn't oceans be present on this planet?

215

Our earth rotates on its axis from east to west. This motion of the planet is what is responsible for our day and night (Figure 11.15).

25. Notice in Figure 11.15 that the time of day for any given area depends on where that area is in relation to the sun. For example, when the sun is directly overhead, it is noon. Given this, when it is 3:00 P.M. in Phoenix, Arizona, what time is it in Bombay, India, halfway around the world?

In addition to rotating, our earth also revolves around the sun (Figure 11.16). One complete revolution takes about 365 ¼ days (one year).

Notice on Figure 11.16 that the earth's axis is not straight up and down. Instead, it is tilted somewhat from the vertical. This tilt (23 ½ degrees from vertical) accounts for the seasons we experience as our earth revolves around the sun.

26. (a) Looking at Figure 11.16, as the earth moves from its position on March 21 to its position on June 21, does the Northern Hemisphere receive more sunlight or less sunlight?

(b) Which hemisphere is receiving the most sunlight on December 21?

ACTIVITY

Holding a globe, walk around a person sitting in the middle of the room. If this person represents the sun, what season of the year is it at different locations around the room?

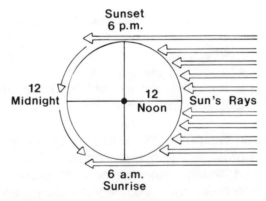

FIGURE 11.15   Rotation of the Earth

216

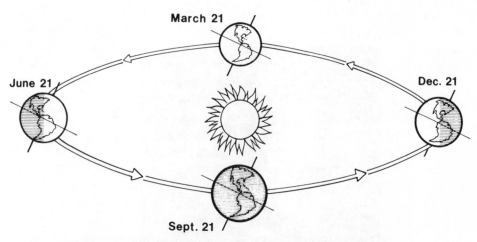

FIGURE 11.16   The Earth Revolving around the Sun

The Moon

Our moon is the only natural satellite of earth. It is slightly smaller than the planet Mercury and it has no atmosphere. This means, of course, that you could not breathe on the moon.

The moon takes approximately four weeks to revolve around the earth. As it does so, it goes through different phases. These phases are shown in Figure 11.17. As you look at this figure, keep in mind that half of the moon is always illuminated by the sun. But, because of our perspective here on earth, we usually see only a part of this illuminated side.

27. When the moon is at first quarter, we see about half of its illuminated side. At what other phase do we see about the same amount of the moon?

28. When the moon is full, we see all of its illuminated side. At what phase do we see little if any of the moon's illuminated side?

29. As seen in Figure 11.17, the surface of the moon is extensively cratered, due to the impact of meteors. Our earth is not cratered like this, in part because our atmosphere burns up most meteors before they strike the surface. Why don't meteors burn up before they strike the moon's surface?

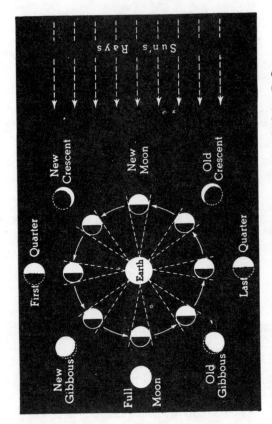

FIGURE 11.17 The Moon (Left: Lick Observatory Photograph. Right: Copyright © 1959 by D.C. Heath and Company)

SUMMARY

Working from the observatory of the House of Science, astronomers have compiled a picture of the universe that is much more encompassing than was ever imagined by people of past ages. In this universe, distances can be measured in terms of light years, with our own Milky Way galaxy being 100,000 light years in diameter. Also in this universe, stars burn their fuel in a process known as nuclear fusion. As they near the end of their lives, some of these stars undergo changes that are sometimes spectacular.

The tools astronomers use include various telescopes and the spectroscope. The spectroscope analyzes light and breaks it apart into dark lines of different wavelengths. When we look at distant galaxies with this instrument, we notice a shifting of these dark lines. This shift is thought to be due to the galaxies moving away from us in the aftermath of the Big Bang.

Our solar system is the part of the heavens we are most familiar with. It contains nine planets, of which our earth is one. Circling the earth is our only natural satellite, the moon.

30. There is much emphasis today on space travel. Telescopes in space, for example, don't have to contend with disturbances in the earth's atmosphere. People who are in space also don't have to contend with gravity. Why is this?

ANSWERS

1. Counterclockwise

2. 200 billion

3. 100,000 light years × 5.8 trillion miles/lt. yr. = 580,000 *trillion* miles

   In metric:

   100,000 light years × 9.3 trillion km/lt. yr. = 930,000 *trillion* km

4. 33 years round trip

5. The light from Andromeda takes two million years to reach us.

6. They would be seeing us as we were two million years ago.

7. About ten billion years

8. The other stars would be too faint to detect individually.

9. The bright sun obscures this starlight.

10. Heavy elements are thought to form only in supernovas.

11. The nucleus is very small in comparison to the distance of the

orbiting electrons. The space in between is just that—empty!

12. They wouldn't actually be able to see it.

13. Prime focus

14. The human eye is not capable of seeing radiation of this wavelength.

15. (a) A telescope
    (b) A prism

16. The indigo portion of the spectrum

17. To the left of where they normally appear

18. (a) B—two units
        C—three units
        D—six units
        E—eight units
    (b) Yes
    (c) Yes

19. 186,000 mi./sec. × .9 = 167,400 mi./sec.

    In metric:

    300,000 km/sec. × .9 = 270,000 km/sec.

20. More black holes would provide more gravity to stop the expansion.

21. Protoplanets

22. Jupiter, Saturn, Uranus, and Neptune

23. The Asteroid belt

24. The water in them would boil away!

25. 3:00 A.M.

26. (a) More sunlight
    (b) The Southern Hemisphere

27. Last Quarter

28. New Moon

29. The moon has no atmosphere.

30. They are weightless.

Chapter 12

EPILOGUE

Now that you have completed your tour of the House of Science, you should be better able to understand the role of science in our present-day world. This role has been a leading one, for it has given us the life styles we currently live. Science has influenced how we care for our bodies, how we travel from place to place, and how we communicate with each other. It has influenced how we conduct business, how we conduct warfare, and, in general, how we view the world. Indeed, as the twentieth century draws to a close, we must in all truthfulness say that it has been the century of science!

As we look to the future, we will find scientists opening up new frontiers of understanding. As they do so, we will be reminded from time to time that the rooms in the House of Science are interconnected. Discoveries made in physics, for example, will help astronomers in their quest to better understand the universe. Likewise, discoveries in chemistry will be useful to biologists as they strive to better understand our living world.

As we look to the future, we must also take with us a knowledge of the past. For example, we must remember how science was used in past ages—both for good and for evil purposes. With this knowledge, we must then strive to ensure that science works for the good in the future world we will inherit.

INDEX